Physics of Meteor Flight in the Atmosphere

PHYSICS OF METEOR FLIGHT IN THE ATMOSPHERE

Ernst J. Öpik
Armagh Observatory
Armagh, Northern Ireland

DOVER PUBLICATIONS, INC.
Mineola, New York

DOVER PHOENIX EDITIONS

Bibliographical Note

This Dover edition, first published in 2004, is an unabridged republication of the work originally published in 1958 by Interscience Publishers, Inc., New York.

International Standard Book Number: 0-486-43885-6

Manufactured in the United States of America
Dover Publications, Inc., 31 East 2nd Street, Mineola, N.Y. 11501

Introductory Note

The booklet has originated from a series of lectures on "The Physics of Meteors and their Cosmic Relationships," delivered in the spring term of 1957 at the Department of Physics of the University of Maryland, during the author's stay there as Visiting Professor.

The present scope is somewhat narrower and covers chiefly the phenomena occurring during the flight of the meteor through the terrestrial atmosphere. The intention is to provide a basis for further research.

The basic units of the cgs system are used throughout unless stated otherwise. Thus, normally lengths will be in centimeters, velocities in cm/sec, and so on.

A number of symbols, listed as Principal Notations at the end of the book, will have always the same meaning and, to save space, will be defined only when they occur for the first time. Other notations will be currently defined in the text.

Contents

Physics of Meteor Flight in the Atmosphere

CHAPTER 1

The Setting

1.1 Problems

The physical theory of meteors considers processes which are taking place during the flight of the meteor through the atmosphere and, for larger bodies, at impact on the earth's surface. The main purpose is to predict the variation of mass, velocity, luminosity, and ionization along the meteor trajectory.

In meteor physics, calculations are sometimes only approximate; however, results to a "close order of magnitude" ($\pm$ 50 %) often suffice.

The depth of penetration of meteors into the atmosphere depends mainly upon their size and velocity. The quantitative difference in atmospheric density may be so great that a qualitative difference in the treatment becomes necessary. As a consequence, the theory of the processes occurring in collisions of meteors with our planet can be subdivided into three major "problems," with transitions between them.

Problem 1: This concerns the case when the "free path" of the air molecules is greater than the linear dimension (radius) of the meteor. The impact momentum and energy are transmitted to the nucleus of the meteor by direct hits of the air molecules. No hydrodynamic cushion or "air cap" is formed. Practically all meteors of the "visual" range belong to this case; they are so small and disintegrate in atmospheric strata of so low a density that the condition of free path is always amply fulfilled (Fig. 1).

One consequence of the lack of shielding is the relatively high coefficient of heat transfer; with this, only a small fraction of the kinetic energy of the meteor is needed to achieve complete vaporization of its substance; the meteor disintegrates before noticeably decelerating. Hence in many cases the problem can be further simplified by assuming the velocity to remain nearly constant.

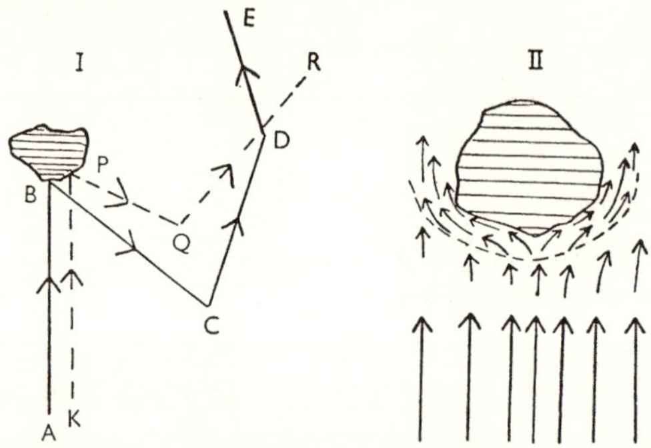

Fig. 1. Motion of air relative to meteoroid. I = Problem 1. The air molecule A hits the meteoroid at B and rebounds along the zig-zag track BCDE, being in collision with other air molecules at C and D; the stretches BC, CD are large as compared with the diameter of the meteoroid.

II = Problem 2. The air stream obviates the obstacle along streamlines shown by arrows. Between the broken line and the meteoroid an air cap, or condensation, is formed. The free path of the molecules is small as compared with the diameter of the meteoroid.

Problem 1a: This is the particular case of micrometeors which are efficiently decelerated in atmospheric strata of so low a density that the temperature of intense vaporization is not attained, and the energy is dissipated through radiation. Micrometeors may be brought to rest without much of ablation, and may in the course of time settle to the ground as particles of dust.

Problem 2: This is the case when the free path of the air molecules is smaller than the linear dimension of the meteor. A hydrodynamic cushion, or air cap, is formed in front of the meteoroid (nucleus) (Fig. 1). The heat transfer is impeded, the more so, the deeper the meteor penetrates into the atmosphere. The aerodynamic resistance, or drag is also smaller than for equal velocity and atmospheric density in Problem 1, because the air cap helps in streamlining the flow of air. Once the meteor has penetrated deep enough into the atmosphere for an air cap to be formed, the protective quality of the latter will favor further penetration. The heat transfer coefficient may decrease by two orders of magnitude, as compared with Problem 1, whereas the atmospheric drag can only decrease by a factor of about 2. Ablation is greatly suppressed, yet deceleration is affected very little. For a sufficiently large initial size, the meteor may be stopped before it has completely disintegrated.

So-called "fireballs" are of a brilliance equal to that of the moon, or brighter, and the meteorite falls belong to this class.

Problem 3: This is the case of collision of a meteor with a dense body, solid or liquid. From the preceding two cases this differs by the meteor and the medium being of comparable density. The conditions are extremely complicated. On account of the high aerodynamic pressures and plastic deformation, the distinction between meteor and medium is practically erased; the propagation of shock waves and shock fronts, instead of the motion of the meteor body, is to be considered. The theory of formation of meteor craters, collisions between meteors, or between asteroids in space, and meteoric skin erosion of space vehicles belong to the scope of Problem 3.

In the present tract we will concern ourselves mainly with the first two problems.

1.2 Historical

The beginnings of a physical theory of meteor phenomena seem to have been first laid down by Öpik in 1922.[1] A relation of general validity between the variable mass and velocity of a meteor was derived. It was established that the intensity of observable radiation from a meteor is proportional to the mass of vapor set free per unit of time, or that brightness can be measured by the rate of ablation. A light curve, or variation of brightness of the meteor along its path, was calculated for an atmosphere of constant density.

Hoppe[2] derived a light curve upon the same principle for motion in an atmosphere of variable density.

An attempt of a detailed physical theory of meteors was made by Lindemann and Dobson in 1923.[3] Unfortunately, it was marred by serious misinterpretations of the laws of physics. Their most serious shortcoming amounted to a neglect of the law of conservation of energy; namely, they calculated the temperature of air in front of the meteor from the trivial adiabatic formula of compression; yet the adiabatic formula is only valid when the compression is slow as compared with the gaskinetic velocity of the molecules. This condition is not fulfilled in the compression of air in front of the moving meteor. In the words of Sparrow,[4] "The use of the adiabatic equation by Lindemann and Dobson is equivalent to the assumption that a velocity of 60 km/sec is small compared to one of 0.5 km/sec." The procedure cannot even be called an approximation, less than 1 per cent of the kinetic energy being accounted for. For this and other reasons their treatment of the meteor problem is unacceptable.

A most important step forward in the physical theory of visible meteors was made in 1926 by Sparrow.[4] He pointed out that these objects (meteors of the visual range) cannot produce an air cap and that momentum and energy transfer are achieved through direct bombardment of the meteor nucleus by air molecules. Sparrow actually laid the foundation of the modern detailed theory of the balance of energy and mass in visible meteors (Problem 1). A further contribution to the meteor theory on rational lines came from H. B. Maris.[5]

A study entitled "Basis of the Physical Theory of Meteor Phenomena" was published by Öpik in 1937.[6] This is an analysis of allegedly all relevant factors, instrumental in the process of ablation and atomization of the meteor substance. No doubt, many a factor may nevertheless have been omitted, such as, e.g., sputtering, or the newly discovered phenomenon of dustballs. However, it seems that the enumeration is reasonably complete from the standpoint of processes occurring in stonelike visible meteors. Together with an earlier paper, titled "Atomic Collisions and Radiation of Meteors,"[7] which contains a sketch of a quantitative semi-empirical theory of the efficiency of ionization and excitation in collisions of slow atoms or ions, this study has served as a basis for further research.

The factors numerically evaluated included aerodynamic drag, coefficient of accommodation, radiation losses, vaporization, fusion, viscosity and spraying of liquid at fusion, stability of liquid drops in flight, role of rotation, aerodynamically induced oscillations and damping of rotation, formation of an air cap, shielding effect of own vapors, heat transfer by conductivity, and impact radiation from vaporized substance.

From this stage, further theoretical progress can be achieved mainly by numerical integrations. The integrations need not be of high numerical accuracy, but of paramount importance is the condition of a realistic approach; this means that no relevant factor should be overlooked or omitted merely for the purpose of mathematical expediency.

Meteor theory has been progressing through the application of the method of trial and error. A provisional set of assumptions is followed by analysis which shows where the assumptions are in need of adjustment; the adjusted set is taken for a new start, and the procedure is repeated until consistent results are obtained.

It may happen that inconsistency cannot be eliminated, which would mean that some of the assumptions are basically wrong. Such was the case with the analysis of the heights of appearance and disappearance of visible meteors;[8,9] the observational results, coupled with atmospheric data obtained from rocket research, could not be reconciled with any rational assumptions as to the process of ablation of a stony meteoroid. The conclusion was reached that the meteors of the visible range, as a rule, are not stony compact bodies, but some kind of loose aggregates of dust particles, "stoneflakes" or "dustballs." Although the "classical" theory of ablation, as it could be called, proved inapplicable, without the help of this theory the correct conclusion could not have been reached.

In other cases, e.g., for micrometeors and for large meteors (fireballs) and meteorites, the assumption of a dense stony (or iron) kernel has proved consistent with the observations.

CHAPTER 2

The Atmosphere

2.1 Meteors as Probes of the Upper Atmosphere

The correct interpretation of meteor phenomena presupposes a knowledge of the structure and composition of the atmosphere to a height of at least 130 km.

On the other hand, it may seem to be possible to derive the properties of the atmosphere from meteor observations when the average density and shape of the meteors are postulated. In such a manner, Whipple[10] calculated atmospheric densities from the deceleration of bright photographic meteors within the altitude range of 60 to 95 km. The results, however, are subject to large systematic errors which are altitude-dependent and, therefore, are of doubtful value.

Discrepancies for meteors of the visual range led to the recognition of dustballs[8,9] showing that our knowledge of meteors is much less certain than that of the atmosphere. The problem is to be inverted and, by taking as a basis the observed structure of the atmosphere, e.g. that of the Rocket Panel,[11] one has to infer the properties of meteors that are compatible with the observations.

2.2 Diffusion and Turbulence

A quarter of a century ago the problem of the structure of the upper atmosphere was bedevilled by the theory of conductive, or diffusive equilibrium (Humphreys, Jeans). In undisturbed gas diffusion the constituents of the atmosphere settle into states of equilibrium which are independent of each other; the partial pressures are then governed by the equation

$$dp/p = -dH/a, \tag{2–1}$$

where

$$a = kT/(\mu g) \tag{2–2}$$

is the scale height. Here p = partial pressure, μ = molecular mass (absolute), k = Boltzmann's constant, g = acceleration of gravity, H = altitude. The pressure of a light gas like hydrogen will decrease more slowly with altitude than that of the heavy constituents and, although relatively insignificant at sea level, hydrogen would predominate in the upper atmosphere (above H = 80 km). This concept proved a mathematical fiction in the true sense of the word. Sparrow,[4] and especially Chapman[12] denied any role to hydrogen. Öpik[6] pointed out that, contrary to observation, atomic lines of more than 4 ev excitation could not appear in the spectra of meteors at velocities below 30 km/sec if hydrogen were the chief constituent; the lack of differentiation of the atmospheric constituents he explained by turbulent mixing.

Some properties of turbulence which play a role in the theory of meteors are outlined below.

Turbulence replaces laminar flow when the Reynolds number,

$$Re = \rho v L/\eta, \tag{2–3}$$

exceeds a certain limit of the order of 1000. Here ρ is the density, η the coefficient of viscosity of the fluid, v an average value of the velocity of flow, and L a characteristic linear dimension (radius of pipe, thickness of layer of flow at right angles to the direction of motion, etc.). $D_g = \eta/\rho$ is the kinematic viscosity, identical with the gaskinetic coefficient of diffusion. With Re = 1000, Eq. 2–3 defines the maximum admissible thickness of a laminar layer (skin layer).

In the free upper atmosphere $L \sim a \sim 1.2 \times 10^6$, $v \sim 10^4$, $\eta = 2 \times 10^{-4}$ (at $+$ 300°C), whence

$$Re \cong 6 \times 10^{13}\rho. \tag{2–4}$$

Hence, for $\rho > 1.7 \times 10^{-11}$ g/cm^3 the Reynolds number indicates spontaneous turbulent flow; this corresponds to an altitude below 130 km.[11] Forced turbulence will extend to even greater altitudes, being caused by the impact of turbulent elements from below.

Turbulence leads to accidental fluctuations of the velocity components; for a normally horizontal flow, such as the winds in the ionosphere are (average wind velocity 95 m/sec in 100–120 km level;[13] 50 m/sec in 80–100 km level;[14] 54 m/sec at $H = 90$ km and 83 m/sec at $H = 98$ km[15]), vertical components arise which are instrumental in mixing the atmosphere through turbulent or eddy diffusion. Although favored by mountains, these vertical motions belong to the very nature of turbulence; there is no need for solid obstacles to cause them—they are produced in the collisions of disorderly moving turbulent "packets" in the free atmosphere.

Measurements with a sensitive anemometer by Öpik (1927–39, unpublished study of diffusion of gas clouds) give an average vertical deviation of wind direction of $\varepsilon = \pm 6°$ even under the smoothing presence of a level surface; being a time average, this implies a semiamplitude in the vertical velocity component equal to

$$u = 2v \sin \varepsilon = 0.2v, \tag{2–5}$$

v being the wind velocity.

In an atmosphere of stable lapse rate an air packet which has acquired a vertical velocity component u will cool from adiabatic expansion while rising; it will be slowed down by a differential acceleration of gravity equal to

$$\Delta g = F_1 g(\nu - \nu_0)x/T \quad (\text{cm/sec}^2), \tag{2–6}$$

where x is the differential height, $\nu = dT/dx$ the lapse rate, ν_0 the adiabatic lapse rate, and F_1 a dimensionless factor which allows for the lateral exchange of heat by second-order turbulence; empirically $F_1 = 0.096$.[16] The work per unit mass is

$$\int_0^x \Delta g \, dx,$$

and the vertical motion stops when this equals $\frac{1}{2}u^2$. Hence the maximum vertical length of path of the air packet becomes

$$x_{\max} = u[F_1 g(\nu - \nu_0)/T]^{-\frac{1}{2}}. \tag{2-7}$$

With $T = 360°K$, $\nu = +0.6 \times 10^{-4}$ deg/cm as for altitudes of 110–140 km,[11] $\nu_0 = -0.8 \times 10^{-4}$, $g = 940$, we have $x_{\max} = 220u$. With $v = 10^4$, $u = 2000$, $x_{\max} = 4.4 \times 10^5$ cm. Allowing for aerodynamic resistance and dissipation of the air packet, the vertical exchange depth for turbulent mixing may be assumed equal to one-half of that, or to $\Delta H = 2.2 \times 10^5$ cm.

The coefficient of turbulent diffusion in the vertical direction, equal to the product of effective length of vertical path times average vertical velocity component, thus becomes

$$D_t = \tfrac{1}{2}u \cdot \Delta H, \tag{2-8}$$

or

$$D_t = 2.2 \times 10^8 \text{ (cm}^2\text{/sec)}. \tag{2-9}$$

The flow of mass in the direction x is given by

$$\rho\left[\left(\frac{\partial z}{\partial x}\right) - \left(\frac{\partial z}{\partial x}\right)_0\right]$$

(g/cm² sec), where $\partial z/\partial x$ is the gradient of partial concentration of an atmospheric constituent, $(dz/\partial x)_0$ the equilibrium value of the gradient. For turbulent diffusion $(\partial z/\partial x)_0 = 0$. For gaseous diffusion $(\partial z/\partial x)_0$ is determined by Eq. 2–1, with the actual local values of p and T.

The isotropic coefficient of gaseous diffusion is given by

$$D_g = \tfrac{1}{3}\lambda_0 \bar{v}, \tag{2-10}$$

where λ_0 is the mean length of path (for 90° deflection in angle), $\bar{v}$ the gaskinetic arithmetical mean velocity.

For an initial density distribution of excess substance, cylindrically symmetrical around a straight axis (as in ionized columns of meteors) and of the form exp $(-x^2/\Delta^2)$, the diffused distribution after the lapse of time t becomes exp $[-x^2/(\Delta^2 + 4Dt)]$ when there is no external supply of the substance; here x is the distance from the axis and D the coefficient of diffusion. When the initial dispersion, Δ, is small, $(3Dt)^{\frac{1}{2}}$ is about the mean (or 50 %-concentration) distance. For one-dimensional diffusion from a plane "wall" with unlimited supply and initial concentration zero at $x \geqq 0$, the relative distribution is determined by

$$\int_{z}^{\infty} e^{-z^2}\, dz,$$

with $z = x/[2(Dt)^{\frac{1}{2}}]$; $(2Dt)^{\frac{1}{2}}$ is the mean-square distance and $(Dt)^{\frac{1}{2}}$ very nearly the 50 % concentration distance.

From the Rocket Panel[11] it appears that $D_g = D_t$ at an altitude of 180 km. Below this altitude turbulent mixing is probably efficient, and the chemical composition of the atmosphere can be assumed constant.

At greater altitudes diffusive equilibrium may become important, and the relative abundance of the light gases, hydrogen and helium, may increase. However, forced turbulent mixing by horizontal flow is not the only mixing agent. Large-scale circulation in the upper atmosphere, with slow ascending and descending currents over wide areas of the globe, may be even more efficient in this respect,[17] and may equalize the composition of the atmosphere to even greater altitudes.

2.3 Structure of the Upper Atmosphere

The data for the atmosphere above 160 km are very uncertain; in the Rocket Panel[11] and in a synoptic review by Bates[18] they are actually based on only two successful rocket ascents.

Observations of meteorites at an early stage of their flight may provide additional evidence for the density gradient of the atmosphere at these heights. This is the case with the great meteorite of January 30, 1868, of Pultusk Poland, which broke up from aerodynamical pressure at a height of 42 km ("Hemmungspunkt") into stony fragments, about 100,000 of which have been recovered.[19] It was noticed by an experienced observer when at a height of 260 km; from that height to 180 km it increased in brightness by an estimated 6 magnitudes, or in a ratio of 250 to 1. This would correspond to a "photometric" scale height of 14.7 km. At these altitudes the luminescence could only have been due to "impact radiation," or to the direct bombardment of the solid surface by air molecules, the heating of the surface being a few degrees only and vaporization not being possible. The impact radiation must be proportional to the air density, and the photometric scale height should correspond to the true scale height. According to Bates,[18] the density ratio over the range from 260 to 180 km is only 23, giving a scale height of 25.5 km. The observation suggests a more rapid decrease of density with altitude; the temperature of the atmosphere in this region may perhaps be lower, about 440° K instead of 760° assumed by Bates,[18] leading to very much lower atmospheric densities at the 260 km level. However, preliminary results from satellites seem to favor a greater scale height.

Table I contains a summary of the structure and composition of the atmosphere based on the Rocket Panel; some new determinations, including those based on X-ray absorption challenge the accuracy of these data, indicating from 2 to 3 times smaller a density between 90 and 140 km.[20] Nevertheless, we agree to use the Rocket Panel densities and other relevant data as a standard of comparison until better data will be available.

An important correction to prevailing views on the constitution of the upper atmosphere may be suggested, according to M. Nicolet.[18] It has been ordinarily assumed that molecular oxygen is efficiently dissociated by solar ultraviolet radiation, so that above $H = 102$ km O_2 will be practically absent;[21] this follows from Chapman's assumption of an effective temperature of 6000° K for solar radiation in the relevant portion of the ultraviolet (< 2300 A). However, measurements made during rocket ascents indicate a temperature of only about 4500° K. The number of quanta available for dissociation of O_2 is thus about 250 times less

TABLE I

Structure of the Atmosphere from Rocket Data[11]
(temperature and molecular weight are adjusted)

H km	Mol. weight	T °K	$\log p$ d/cm²	$-\log \rho$ g/cm³	a km	g cm/sec²	λ_0 cm	D_g cm²/sec
5	28.97	276.8	5.742	3.140	7.83	978.0	1.2×10^{-5}	0.16
10	28.97	230.8	5.446	3.375	6.78	976.5	2.1×10^{-5}	0.27
20	28.97	212.8	4.755	4.030	6.28	973.4	9.7×10^{-5}	1.1
30	28.97	231.7	4.095	4.728	6.85	970.4	4.8×10^{-4}	5.8
40	28.97	262.5	3.497	5.380	7.79	967.3	2.2×10^{-3}	28
50	28.97	270.8	2.955	5.936	8.06	964.3	7.8×10^{-3}	102
60	28.97	252.8	2.404	6.457	7.55	961.3	2.6×10^{-2}	330
70	28.97	218.0	1.784	7.012	6.53	958.4	9.3×10^{-2}	1.1×10^{3}
80	28.97	205.0	1.094	7.676	6.16	955.4	0.43	4.9×10^{3}
90	28.97	217.0	0.405	8.389	6.54	952.4	2.1	2.5×10^{4}
100	28.97	240.0	−0.227	9.065	7.26	949.5	9.5	1.2×10^{5}
110	28.97	270.0	−0.794	9.684	8.19	946.6	38	5.0×10^{5}
120	28.97	330.0	−1.273	10.249	10.04	943.6	130	1.9×10^{6}
130	27.52	370.5	−1.670	10.720	11.90	940.7	370	5.8×10^{6}
140	26.22	404.6	−2.009	11.119	13.68	937.9	870	1.5×10^{7}
150	25.03	423.7	−2.308	11.468	15.44	935.0	1800	3.2×10^{7}
160	23.95	463.0	−2.574	11.781	17.24	932.1	3600	6.8×10^{7}
170	23.1	494	−2.813	12.062	19.11	929.3	6100	1.2×10^{8}
180	22.3	521	−3.030	12.318	20.97	926.4	1.0×10^{4}	2.1×10^{8}
190	21.6	548	−3.228	12.552	22.84	923.6	1.8×10^{4}	3.9×10^{8}
200	21.0	575	−3.411	12.768	24.70	920.8	3.0×10^{4}	6.7×10^{8}
210	20.3	596	−3.580	12.968	26.57	918.0	5.1×10^{4}	1.2×10^{9}
220	19.8	620	−3.738	13.154	28.43	915.2	8.7×10^{4}	2.1×10^{9}

than formerly assumed, the lifetime of the oxygen molecule being around 10 days, instead of about one hour as was formerly estimated; thus, the degree of dissociation of O_2 must be very much smaller. This is a very important point, in view of the role O_2 plays in the recombination of ionized columns in the wake of meteors.[9] Owing to its low ionization potential, it takes over the charge from the other atmospheric constituents (N_2^+, N^+, O^+) and doubly ionized metal atoms produced in atomic collisions, and favors a rapid rate of recombination in the ionized column (dissociative recombination).

In Table I it has been conventionally assumed that the transition from O_2 to O takes place at the 160-km level, instead of at 102 km.

The coefficient of turbulent diffusion in the free atmosphere and in a vertical direction can be assumed constant as given by Eq. 2–9, provided exchange is considered between layers of at least several km thickness, or well in excess of ΔH (Section 2.2); in a horizontal direction the coefficient will be greater.

For small-scale phenomena, such as in the wake columns of meteors, the coefficient of turbulent diffusion is estimated upon quite different principles.

TABLE II
List of Atmospheric Constituents

Molecule	Ionization pot., ev	Dissociation en., ev
N_2	15.57	9.76
N	14.54	—
NO	9.5	6.50
O_2	12.21	5.08
O	13.614	—
H_2	15.42	4.48
H	13.595	—
A	15.76	—
Na	5.14	—
N_2^+	—	8.72
O_2^+	—	6.48
H_2^+	—	2.65

In Table II the ionization and dissociation energies of the most important atmospheric constituents are listed.

The number density of hydrogen (atomic) at 100 km[22] is from 10^{-5} to 10^{-4}, and that of helium 5×10^{-6}.

2.4 Recombination

The coefficient of recombination of room-temperature electrons with atomic ions (radiative recombination) is $\chi \cong 10^{-11}$ cm^3/sec, and that of dissociative recombination[23] with molecular ions about 10^{-7}; $\chi \sim T^{-0.75}$ approximately (the negative exponent increases with temperature from $-\frac{1}{2}$ to -1 and over). In the ionosphere, intermediate values of the effective coefficient of recombination are observed; these can be explained by complicated processes of exchange of charge; the effective value of χ there chiefly depends upon the concentration of O_2, the constituent with the lowest ionization potential, O_2^+ being of high activity in recombination as a molecular ion.

The probability of capture per second of an electron by an ion is $\chi_i N_e$, where χ_i is the recombination coefficient of the particular ion, and N_e the electron density. The number of captures per cm^3 and sec is

$$\frac{dN_e}{dt} = -N_e \Sigma_i \chi_i N_i, \tag{2–11}$$

the sum being taken over all existing ions. Usually $\Sigma_i N_i = N_e$, and the effective coefficient of recombination is $\bar{\chi} = \Sigma_i \chi_i N_i / N_e$, whence

$$\frac{dN_e}{dt} = -\bar{\chi} N_e^2. \tag{2–12}$$

For constant $\bar{\chi} = \chi$ and initial electron density N_0, integration of Eq. 2–12 yields the electron density after time t in unilateral decay (i.e., when sources of ionization are absent):

$$N_e = (N_0^{-1} + \chi t)^{-1}. \tag{2–13}$$

The half life of decay, when $N_e = \frac{1}{2} N_0$, follows from Eq. 2–13 as

$$t_0 = (\chi N_0)^{-1}. \tag{2–14}$$

More complicated expressions result when χ is the resultant of heterogeneous components.

The role of negative ions in meteor physics is negligible.

The two-body (radiative) recombination of atoms into diatomic molecules is also too slow to play a role in meteor physics.

2.5 Sound

The absence of sound from meteors appearing in the upper atmosphere is primarily due to gaskinetic absorption. This reduces the intensity of sound in a ratio of

$$I/I_0 = e^{-bx}, \tag{2-15}$$

where x is the distance, and b the coefficient of absorption, approximately given by

$$b = 4\lambda_0/L^2, \tag{2-16}$$

L being the wavelength of sound. The absorption increases with decreasing density of the atmosphere and decreasing wavelength. For a very low tone of $L = 10^3$ cm, the intensity will be reduced 20000 times over a path of 2.5 kilometers at $H = 100$ km ($\lambda_0 = 9.5$ cm). The low tone of meteor detonations can be explained by selective absorption which varies as L^{-2}.

CHAPTER 3

Classification and Physico-Chemical Properties of Meteors

3.1 Meteor Classification and Nomenclature

"Meteors" is the common name for all bodies entering the terrestrial atmosphere from interplanetary space, irrespective of size and origin. Their behavior is determined by their size, structure, and velocity; the latter cannot be less than 11.2 km/sec (escape velocity) and, for the majority of visible meteors, ranges from 20 to 70 km/sec.

The term "meteors" (also "ordinary" or "visible" meteors) in a more narrow sense refers to those bodies which are too small to penetrate the atmosphere and which become observable only through their luminous paths in the upper atmosphere; in the order of increasing size they are telescopic meteors, visual meteors, photographic meteors, and fireballs. The limits of size are roughly from 0.03 cm to 10 cm equivalent radius.

The term "meteorites" in a narrow sense refers to those remnants of meteoric bodies which have penetrated the atmosphere and are found on the ground. In a wider sense this term covers all large meteors capable of yielding recoverable fragments, being thus the next step in the meteor hierarchy after the fireballs. Their preatmospheric size should exceed 10 cm equivalent radius.

"Micrometeors" is a term used for particles smaller than 0.03 cm equivalent radius; the radii appear to be most frequent around 10^{-3} cm, and reach down to 10^{-4} cm at least. They are too small to be observed with optical means. While forming the bulk of the zodiacal-light cloud, their mass per volume of space greatly exceeds that of all other meteors taken together. Most of them possess small relative velocities with respect to the earth and enter our atmosphere with speeds of the order of 12 km/sec. Because of small velocity and small size they dissipate most of their energy through radiation, with little vaporization, and are stopped in their flight with the greater portion of their mass preserved; in all appearance, those metallic micrometeors which have become melted during flight are collected from the deep-sea clay as the well-known "cosmic spherules." They could be called "micrometeorites" in this case.

Strictly speaking, all "ordinary" meteors are likely to yield a fused residual kernel; in most cases, however, these kernels represent but a very small fraction of the original mass—the smaller, the greater the velocity. The micrometeorites produced by ordinary visible meteors are very much less numerous than those due to micrometeors, and it appears to be difficult to devise means for their separate detection.

3.2 Compact Meteors and Dustballs; Artificial Meteors

Not so long ago meteors irrespective of size were thought of as compact stones or iron chunks, of irregular shape but firmly held together by the forces of rigidity. The recovered specimens, the meteorites, favored such a concept. It was admitted that some stony meteorites may have been loosely-bound conglomerates but, as, e.g., was actually the case with the Pultusk meteorites, each of the components was believed to be a true, compact stone.

However, with the better knowledge of the upper atmosphere, it became apparent that observations of visual meteors, and of some belonging to the photographic range, could not be interpreted in such terms; these meteors disappear at great altitudes where the density and mass of the atmosphere is insufficient to cause their complete vaporization (or even fusion). The concept of "dustballs" had to be introduced. They are thought to be loosely-bound agglomerates, not necessarily spherical, of smaller dust particles. The dustballs collapse under an aerodynamic drag of about 10^4 d/cm^2. They are thus converted into a dust cloud in which vaporization is favored by the greatly increased area (Fig. 2).

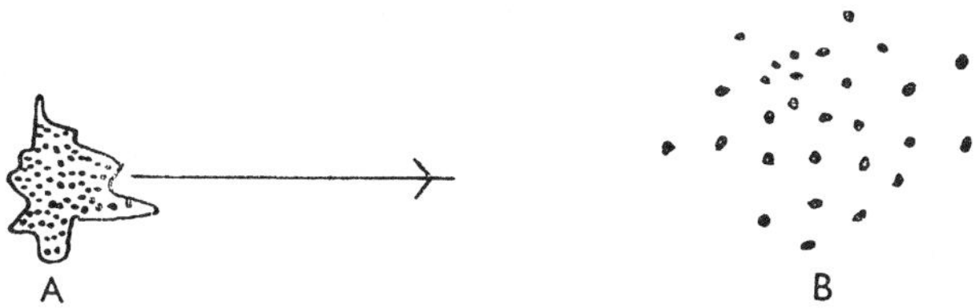

Fig. 2. A dustball (A) decomposes into a cluster of dustgrains.

In contrast with these natural bodies, ultraspeed pellets produced by the explosion of shaped charges may be called "artificial meteors."

Whereas the radiation and ablation of natural meteors takes place in proper altitudes prescribed by a kind of natural selection and depending upon size and velocity, artificial meteors can be released in any surroundings and may therefore lead to phenomena which are not characteristic of natural meteors of equal size.

3.3 Classification of Phenomena

The *meteoroid*, or *nucleus*, is the solid or liquid kernel of the meteor. In the case of dustballs the term becomes ambiguous; the dustball meteoroid breaks up into individual meteoroids or grains.

The jet of vaporized atoms emerging from the meteoroid, mixed with air, will be called *coma* when not yet completely decelerated, i.e., in so far as these elements have retained a considerable fraction of their original kinetic energy of forward motion. In the coma the main dissipation of the kinetic energy of ordinary meteors takes place; it is also the seat of impact radiation, the main source of luminosity of the visible meteors (Fig. 3).

At a sufficient distance behind the meteoroid the translational velocity of the coma is decelerated to well below the mean molecular velocity of the surrounding atmosphere; it is then converted into the *wake train,* in the shape of a column lagging behind the meteoroid; this is identical with the *ionized column* of radar observers. The wake train expands from turbulent and gas diffusion, assuming a paraboloid shape. Except for the first 10^{-3} sec after the passage of the meteoroid, the column is so narrow that it can be closely approximated to a cylinder (with irregularities due to turbulence).

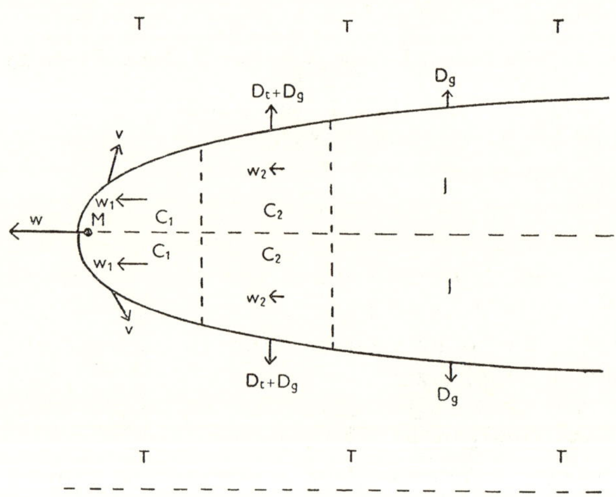

Fig. 3. Meteor phenomena. M = meteoroid (compact, or cluster of grains) with velocity w; the paraboloid is the envelope of expanding meteoric materials. C_1 = front portion of coma; w_1 = translational velocity; v = expansion velocity (direct penetration + expansion of the vapor-air cloud). C_2 = rear portion of coma; w_2 = decreased translational velocity; $D_t + D_g$ = expansion by eddy and gas diffusion. J = wake train; D_g = expansion by gas diffusion. T = persistent train.

During the downward flight the density of the atmosphere and, thus, the intensity of vaporization increases, while the surface area of the meteor decreases from ablation. The luminosity of the meteor, which is a product of the two, varies thus according to a skew *light curve*, with the maximum nearer to the end, as a rule.

The *persistent train*, called also simply the *train*, is observed as a band of light along portions of the trails of bright and fast meteors, after the meteors themselves have disappeared. Its diameter is of the order of from 0.1 to 1 km, and its time of visibility is from a few seconds to several minutes, and more. The train is often confused with the *trail*, which denotes the luminous path of the meteor.

3.4 The Train

The almost instantaneous appearance of the train after the passage of the meteor, at a finite distance on both sides of the trail, shows that it is chiefly caused by radiation, as has been suggested by Trowbridge.[24] Radiation of less than 1000 A ionizes $O_2 \rightarrow O_2^+$ and $N_2 \rightarrow N_2^+$, the latter giving up its charge to O_2^+ almost immediately. In dissociative recombination, $O_2^+ + e \rightarrow O' + O''$, where one or both resulting oxygen atoms may be in an excited state (^1S, 4.17 ev; ^1D, 1.96 ev). There is a residual kinetic energy either of 0.50, 1.48, 1.60, or 2.58 ev per atom which, however, is apparently unable to de-excite efficiently the metastable states in collisions with other molecules. Indeed, various experiments[25] show that the forbidden oxygen line 5577 A ($^1S \rightarrow {}^1D$) is not suppressed by thermal collisions with A and N_2 molecules even at atmospheric pressure, i.e., at 6×10^9 gaskinetic collisions per second; evidently, the de-excitation requires high energies of collision, uncommon at room temperature; the experiments would indicate a lower limit of at least 0.5 ev for the kinetic energy of de-exciting collisions; in the meteor coma, however, the line will be suppressed.[26]

Radiation can be expected from the train in the forbidden lines of OI, 5577 A ($^1S \rightarrow {}^1D$), 6300 + 6363 A ($^1D \rightarrow {}^3P$), and 2972 A ($^1S \rightarrow {}^3P$).

Spread from the wake train of some luminous substance working as a catalyst could be another factor; this, however, could be only a delayed secondary effect, unable to explain the sudden appearance of the train at a finite distance from the meteor wake.

The strong dependence on velocity also suggests some high-energy quantum processes; according to Plavec,[27] the percentage of occurrence of long-enduring trains is in a linear relationship to velocity, % trains $= 1.5v - 45$, v being expressed in km/sec. According to this, below $v = 30$ km/sec trains are not expected to appear.

The absorption cross section of O_2 or N_2 at the ionization limit is of the order of 10^{-17} cm^2, and 10^{-18} is perhaps an overall average for greater radiation frequencies. With $N = 3 \times 10^{13}$ cm^{-3} at $H = 100$ km the average depth of penetration of the ionizing radiation would be about 0.3 km, which corresponds to an initial diameter of the train of 0.6 km, in agreement with observations.

Meteors which lead to trains are of a mass from 0.1 to 1 g, or of a kinetic energy range from 2×10^6 to 2×10^7 erg per cm of path. There are about 4.5×10^8 ions created in the wake train per erg of kinetic energy.[9] Assuming the number of ionizing ultraviolet quanta to be one-tenth of this, the number density of O_2^+ ions within a column of 3×10^4 cm radius becomes from 3×10^4 to 3×10^5 per cm^3. With $\chi = 10^{-7}$ for dissociative recombination, the half-life of decay of the ionization in the persistent train during its initial phase becomes, according to Eq. 2–14, from 300 to 30 seconds, which is of the right order of magnitude.

It remains to obtain a spectroscopic proof of the explanation by looking for the oxygen lines 5577, 6300 + 6363, and 2972 A, a proof which is still wanting.

3.5 Relative Abundance of Meteoric Iron and Stone

There are two principal types of meteoritic matter, iron and stone; between the pure types transition forms occur, the more common being stone with finely divided nickel-iron inclusions, of a low total iron content; less common are mesosiderites, with coarse metal inclusions and a high percentage of Fe, and pallasites, consisting of a nickel-iron matrix with stone (olivine) inclusions of 0.5–2 cm diameter.

The apparent abundance of the two main types, as

judged by their occurrence in museums, is misleading. The prevalence of iron meteorites in collections is explained by the ease with which they are recognized, and by their very much greater resistance to weathering, as compared with stone meteorites. Most stone meteorites in museums belong to observed or recent falls, whereas most of the iron meteorites have fallen in prehistoric times.

There is a remarkable preference of iron meteorite finds for the Western hemisphere; this may be explained by the denser population and the knowledge of the use of iron in the Old World, where all bits of the precious material were eagerly removed long before our museums began collecting them.

The true ratio of meteoric iron to stone can be estimated only under similar conditions of selection, which means that only witnessed and recorded falls should be counted. It appears that the statistics of meteorite falls in the U.S. is best suited for this purpose. Henderson's list of witnessed meteorite falls in the United States[28] names 91 stone falls of a total weight of 3875 kg, and 7 iron falls totalling 69 kg. The mass of metallic iron meteorites amounts thus to only 1.75 per cent of the total; even this may be systematically too high, because iron is better preserved in flight and more readily recovered after a fall than stone.

This conclusion need not necessarily hold for other classes of meteoric matter. A warning against generalizations comes from the consideration of the probable abundance of metallic iron in the earth. The earth's liquid iron core would imply 32.3 per cent of the earth's mass in metallic iron phase; this is very much greater than in meteorites. It is believed that meteorites are fragments of small planets broken up in collisions,[29] and stringent reasons are given for the earth's core being metallic iron, and not a metallic modification of silicates.[29] We have to conclude that the parent bodies of meteorites either contained much less iron phase than the earth, or that their iron cores did not yield many small fragments, remaining either unbroken, or split into a few large pieces. The frequent occurrence of iron fragments around the large meteor craters on earth would seem to support this suggestion, indicating that iron is more common among the crater-producing meteorites.

For micrometeors existing data point also to a very small proportion of the metallic phase, about 0.06 per cent of the total mass of cosmic dust.[30] Furthermore, the photographed spectra of meteors show that stony meteors are in the majority.

The conclusion is that the stony substance greatly prevails among all classes of meteors. The major contribution of metallic phase seems to come from metallic inclusions in stone, whereas the role of pure iron meteors is almost negligible.

3.6 Physical Properties of Iron

Appendix 1 contains a summary of the physical characteristics of iron, the knowledge of which may be required under the various aspects of meteor theory. Only a few comments may be added to this.

The intensity of evaporation from an exposed surface is given by

$$\xi = \tfrac{1}{2}\rho\bar{v}_x, \qquad (3\text{–}1)$$

where ρ = vapor density at saturation pressure, $\bar{v}_x$ = arithmetic mean component of molecular velocity normal to the surface, and the factor $\frac{1}{2}$ takes account of the one-sided flow. This can be reduced to

$$\xi = 4.36\times 10^{-5}\,pm^{\frac{1}{2}}\,T^{-\frac{1}{2}}\ \ (\mathrm{g/cm^2sec}), \qquad (3\text{–}2)$$

where m = molecular weight.

There is very little gas in iron meteorites; they contain from 4 to 33 p.p.m. hydrogen (Canyon Diablo 4.9; Cape York 25.4), less abundant in deuterium than terrestrial ocean water.[31] The helium content is from 0 to 10×10^{-6} cm^3/g of iron.

3.7 Physical Properties of Stone

Our knowledge of the properties of meteoric stone is less complete than that of iron. Nevertheless, a comparison with terrestrial substances helps to fill the gaps. Appendix 2 contains the data for stone.

The minerals of meteoritic stone partly decompose with heating; thus, hematite (Fe_2O_3) dissociates itself from pyroxene at 1250° K, 300° below the temperature of complete melting. The minerals become soft well before melting; the steep decrease in viscosity with rising temperature of the melt is probably also due to breaking up of the molecules into simpler constituents.

At the temperature of vaporization the process of decomposition must have gone very far; the vapors when leaving the surface will contain only certain simple molecules as enumerated in Appendix 2. Vaporization of the stony substance corresponds to a lattice energy level of 2.8–4.2 ev; compounds whose energy of dissociation exceeds this level, such as SiO_2, MgO, Al_2O_3, O_2, will survive vaporization and will be present in the vapors; in others, of a lower energy of dissociation, such as FeO, Fe_2O_3, probably S_2, will decompose.

The relative abundance of a molecule (n_0) and its dissociated components (n_d) is approximately given by the ratio of the Boltzmann factors, or by $n_d/n_0 = \exp_{10}[5040(i_0 - i_d)/T]$, where i_0 is the lattice energy of the molecule, i_d its dissociation energy (ev). Thus, for $SiO_2 \rightarrow SiO + O$, $i_0 = 2.8$, $i_d = 7.0$, and at $T = 2100°$ K (effective temperature of vaporization of stony meteors) $n_d/n_0 \cong 10^{-10}$; the dissociated fraction of SiO_2 will be negligible. For MgO the dissociated fraction will be of the order of $n_d/n_0 \cong 10^{-3}$.

Thus, the temperature of vaporization of meteoric stone, of the order of 2100° K and corresponding to the lattice energy, is too low for appreciable dissociation of SiO_2, MgO, O_2. Therefore, the procedure adopted by A. F. Cook,[64] who identifies thermal vaporization with complete dissociation into atoms, is incorrect and leads to a value for the heat of vaporization of more than three times the probable value.

It is true that the constituents of meteor vapors, after leaving the surface of the meteoroid, will be dissociated in collisions with air molecules; also, sputtering leads normally to the ejection of atoms, not molecules. However, the first phenomenon is subsequent to, and therefore in no connection whatsoever with, the process of meteor ablation, whereas sputtering is insignificant (except in fast and very small micrometeors), as compared with thermal evaporation. There is no justification for Cook's procedure.

The stony meteorites in museums are in a number of cases fragments of larger bodies, broken up by aerodynamic pressure. Large stones reach the denser layers of the atmosphere with practically undiminished velocity; as a rule, they are then breaking up into fragments at altitudes from 4 to 23 km ("Hemmungspunkt"); with velocities from 12 to 30 km/sec (as for objects predominantly in direct orbits) this would indicate an aerodynamic pressure per unit cross section $p_s = \frac{1}{2}\rho w^2 \cong 1.7 \times 10^8$ d/cm^2 as the destruction limit. The figure is by an order of magnitude lower than the compressive strength of basalt and granite, and comes near the lowest figures for sandstone and limestone.

It is probable that shearing stresses caused by aerodynamic drag on an asymmetrical body are the main factor in the breakup. For a prismatic body with an equilateral triangular base moving at right angles to a side the limit of rupture by shear is at an aerodynamic pressure of $p_s = p_r \cong 2\sqrt{3}s'$, where s' is the shearing strength. With $s' = 10^8$ d/cm^2 as an average for terrestrial sandstone, $p_r = 3.5 \times 10^8$ obtains, about the double of the estimate for meteorites. Meteoric stone appears to possess only about one-half the average strength of sandstone.

The Pultusk meteor, $w = 30$ km/sec,[19, 32] broke up at an exceptional altitude of $H = 42$ km, at $p_r = 1.5 \times 10^7$ which is 11 times smaller than the average. It is also the only case of a meteorite with well observed hyperbolic velocity, i.e., of one that arrived from interstellar space. The coincidence of the two exceptional characteristics may be significant. The Pultusk meteorite was still very much stronger than the dustballs, for which $p_r \cong 10^4$ follows from observations.

It is generally believed that various gases are occluded in stony meteorites. The results of former analyses have been subject to doubt by Wahler.[33] He concludes that the announced presence of H_2, CO, N_2, CH_4 is partly explained by diffusion (of H_2 and N_2) from the Bunsen flame used in the experiments, and partly by chemical reactions or dissociation of the gases present.

For the Iowa county stone meteorites Wright (1875/76) found a total volume of gas (at N.T.P.) equal to 2.50 times that of the meteorite. Wahler suggests that originally meteorites contain only CO_2 and H_2O.

Dustballs, to which all cometary meteors belong, are considered to be remains of a more compact material—of cosmic dust cemented by "ices," i.e., volatile substances like CO_2, H_2O, CH_4, NH_3, C_2N_2, which have condensed at the low temperatures of outer space ($< 50°$ K); this concept is the basis of Whipple's comet model.[34] The ices rapidly evaporate in regions of the solar system inside Jupiter's orbit, leaving behind the dustball skeletons. The time of evaporation at 1 astron. unit distance from the sun will be only a few hours for a diameter of 1 cm, i.e., for dustballs of fireball luminosity (-3 mag at 70 km/sec velocity). It is therefore inconceivable that any traces of volatile substances could be retained on the surface of meteors and small meteorites entering the earth's atmosphere.

3.8 Meteor Spectra

At present the number of meteor spectra recorded all over the world has passed the 214 mark. The principal share in their registration and interpretation belongs to P. M. Millman.[35,36] Although in many respects his interpretation is too much linked with the concepts of thermodynamic equilibrium (effective temperature, excitation level) which do not apply in the case of meteors, the basic conclusions are sound.

Meteor spectra are line spectra, due to impact radiation from atomic collisions; traces of an apparently continuous background are discernible, probably due to the merging of a great number of faint lines, not resolvable at the low dispersion of the spectrograms; band emission is also partly suspected.

Some 95 multiplets have been identified by Millman in meteor spectra; because of the low dispersion, as a rule, single lines could not be observed, photometric centres of gravity of the multiplets being used instead. The identified atoms and ions are (subdivided into three groups, approximately in the order of prominence): FeI, MgI, MgII, CaI, CaII, NaI, SiI, SiII; NiI, MnI, CrI, AlI, FeII; HI, NI, OI. There are indications of the presence of the first and second positive groups of bands of $N_2[B^3\Pi_g \rightarrow A^3\Sigma_u^+$ (1.2 volts) in the red, and $C^3\Pi_u \rightarrow B^3\Pi_g$ (3.6 volts) in the violet].[37] The apparent composition revealed by meteor spectra is that of a typical stony meteorite whose evaporated molecules are broken up into atoms in collision with the air molecules. HI belongs apparently to the gases (H_2O) occluded in the meteor grains—the amount of hydrogen in the atmosphere is too small to account for the good visibility of Hα in a Perseid spectrum;[36] NI and N_2 belong doubtless to the atmosphere.

In fast meteors, such as those of the Perseid and Leonid showers, the H (3967 A) and K (3934 A) lines of ionized calcium are especially strong; in other meteors, chiefly of low velocity, they are absent. On this basis Millman originally proposed a classification into two groups, "Y" (with CaII), and "Z" (without CaII). Calcium is virtually absent in iron meteorites, and it appeared at first that a criterion distinguishing between stone and iron meteors has been found. As in the case of stellar spectra, the expectation did not come true: the difference is principally due to the degree of excitation as determined by velocity, and not to chemical composition.

CHAPTER 4

Meteoroid Energy Transfer

4.1 Geometry

Meteor theory contains a source of error depending upon the unknown shape of the meteoroid, except when the latter has become spherical by melting. Certain plausible assumptions can be made in this respect for order-of-magnitude calculations.

Let the *equivalent radius*, r, of a body be defined by

$$M = (4/3)\pi r^3 \delta, \tag{4-1}$$

where M is the mass and δ the average density of its compact material. In what follows, r will be referred to simply as the radius.

The *shape parameter* will be defined through

$$B = \frac{rS}{V}, \tag{4-2}$$

where S = surface area as specially defined, V = compact volume. For a compact sphere $B = 3$; this is the minimum value B can assume. Evidently

$$S = \tfrac{1}{3}B \cdot 4\pi r^2. \tag{4-3}$$

The ratio of mass to area is

$$M/S = r\delta/B, \tag{4-4}$$

and the ratio of volume to area, to be called the average depth, is

$$V/S = r/B. \tag{4-5}$$

In most problems S is the area of the minimum convex envelope of the body.

For a cube of side b, $r = (0.75/\pi)^{1/3}b = 0.6203b$ and $B = 3.721$. For a rectangular prism with sides c, $\sqrt{(cb)}$, b, and $c/b = k$, $r = 0.620bk^{\frac{1}{2}}$ and $B = 1.241\ (k^{\frac{1}{2}} + k^{-\frac{1}{2}} + 1)$ (case I). For a rectangular prism with sides c, c, b, and $k = c/b$, $r = 0.620bk^{2/3}$ and $B = 1.241\ (k^{2/3} + 2k^{-1/3})$ (case II).

Assuming a certain, not very "unusual" relative frequency of the elongation ratios (k), distributions of the shape parameter are found as given in Table III.

TABLE III

Example of Distribution of Shape Parameters

k	Assumed relative frequency	B, case I	B, case II
1/8	2	5.18	5.27
1/4	5	4.34	4.44
1/2	10	3.87	3.91
1	10	3.72	3.72
2	10	3.87	3.95
4	5	4.34	4.69
8	2	5.18	6.20
Average		4.06	4.19
Mean arith. deviation		±0.33	±0.45

For lack of experimental data, these two distributions can be regarded as typical. For very irregular fragments the average value of the parameter B may be somewhat greater, perhaps around a value of 5.

If σ is the cross section of the body with respect to a certain *direction of incidence*,

$$C = \sigma/S \tag{4–6}$$

is the aspect or *contour parameter* for the given direction. Generally $\frac{1}{2} \geqq C \geqq 0$. For a plane thin leaf at normal incidence $C = \frac{1}{2}$, at tangential incidence $C = 0$. For a sphere $C = \frac{1}{4}$; for a cube in orthogonal direction $C = 1/6$, in diagonal direction $\sqrt{2}/4$.

Consider an element of surface dS whose normal makes

an angle α with the direction of incidence; α is the angle of incidence. For *random orientation* the probability of the angle to be between α and $\alpha + d\alpha$ is $\frac{1}{2}\sin\alpha\, d\alpha$. Hence the average value of a function $F = F(\alpha)$ for random orientation is

$$\bar{F} = \frac{1}{2}\int_0^{\pi} F \sin\alpha\, d\alpha. \tag{4–7}$$

Quite often F is "one-sided", such that, for $\alpha > \pi/2$, $F = 0$, which means that F is completely screened by the meteoroid. In such a case in Eq. 4–7 the limits of integration are from 0 to $\pi/2$ only.

We note that Eq. 4–7 is also the expression for averaging F over a *spherical surface*. Thus, the average directional parameters of an arbitrary element of surface in random orientation are identical with those of a sphere.

The contour parameter of an element of surface, $C = d\sigma/dS = \cos\alpha = F$, when one-sided (e.g., for insolation, drag, kinetic energy transfer), has an average value for random orientation as for a sphere,

$$\bar{C} = \tfrac{1}{4}, \qquad \text{or } \bar{\sigma} = \tfrac{1}{4}S. \tag{4–8}$$

The average ratio of mass to cross section (load per unit area) according to Eqs. 4–4 and 4–8 is

$$M/\bar{\sigma} = 4r\delta/B. \tag{4–9}$$

The average thickness, or the ratio of volume to cross section, is

$$V/\bar{\sigma} = 4r/B. \tag{4–10}$$

4.2 Rotation

Rotation has a smoothing effect on ablation when its period, P, is small as compared with the time scale of the meteor, $a \sec Z/w$ (a = scale height, Z = zenith angle of incidence, w = velocity), or $P \ll 6 \times 10^5/w$ (sec). With $P = 2\pi r/u$ (u = equatorial velocity of rotation), the limiting radius for rotational equalization becomes

$$r < 10^5 u/w \text{ (cm)}. \tag{4–11}$$

Rotation is also important in determining the type of ablation (e.g., spraying of the melt). In the following, some estimates of rotation of interplanetary particles are made which replace older estimates.[38]

Meteorites, believed to be fragments of minor planets destroyed in collisions,[29] are expected to carry a not negligible rotational momentum as a relic of the catastrophe. The maximum admissible "equatorial" rotational velocity of a stony body is about $u = 2500$ cm/sec calculated from

$$u \leqq (s_t/\delta)^{\frac{1}{2}}, \tag{4–12}$$

where $s_t = 2 \times 10^7 =$ tensile strength, $\delta = 3.4 =$ density. The observed velocities of rotation of asteroids are of this order of magnitude. With this and (4–11), the limit of equalization is $r < 100$ cm for slow, and < 25 cm for fast meteorites.

Drag forces in interplanetary space (Poynting-Robertson Effect, corpuscular drag) would slow down rotation in the same proportion as the orbital momentum is decreased. If meteorites arrive from the asteroidal region, the rotational velocities, and the limiting radii will be made only 1.5–2 times smaller than estimated.

Unlike the meteoritic fragments, dustballs are not supposed to have originated in an explosion and may possess slow rotation. Yet their fragments, with $r < 0.03$ cm, acquire rotation during fragmentation. Under a pressure of 10^4 d/cm^2 the grains will acquire a velocity dispersion of the order of 130 cm/sec; their velocity of rotation, caused by the friction of adjacent grains before they separate, will be not less than one-tenth of that, or 13 cm/sec. This gives a rotation of the order of over 70 revolutions per second, sufficiently fast for rotational equalization during the lifetime of the grain (0.3 to 1.0 sec).

The micrometeors of zodiacal dust will acquire rotational velocities in mutual collisions. For $r = 0.03$ cm there is at least one collision in 10 years with another zodiacal particle of 10^{-3} cm radius; with 5 km/sec relative velocity this would give a velocity of rotation of about 10–20 cm/sec, or 50–100 revolutions per second from one collision only.

The conclusion is that all meteors smaller than 25–100 cm radius are expected to possess rotation, fast enough to smooth out to a great extent the irregularities of contour or aspect.

The rotation of a meteor may be modified during its flight through the atmosphere. In the case of a nearly symmetrical body, drag will lead to a relative decrease in angular velocity, of the same order of magnitude as that of translational velocity. For small meteors the decrease is insignificant, for meteorites and micrometeors considerable.

On the other hand, asymmetrical bodies, in which the resultant of aerodynamic pressure does not pass through the center of mass, will experience the action of a variable couple influencing rotation. For "contour symmetry," when the aerodynamic pressure on a given contour equals that after rotation by 180°, acceleration of rotation during one-half of the period will alternate with deceleration during the other half. The angular velocity will fluctuate with increasing amplitude, owing to increasing drag as the meteor penetrates into deeper layers of the atmosphere. At a certain depth the negative (decreasing) branch may cross the $\omega = 0$ line (ω = angular velocity), after which rotation ceases and is replaced by oscillation with decreasing amplitude around a certain more streamlined configuration ("tail" turned towards the rear). It can be shown that, for meteors in "normal environment" (cf. Section 5.2), the upper limit of angular velocity of rotation which can be effectively stopped and converted into oscillation before the meteor disintegrates is

$$2\pi/P_0 = \omega_0 \leqq (kh \cos Z)^{\frac{1}{2}} (2\gamma ar)^{-\frac{1}{2}}. \qquad (4\text{–}13)$$

Here h = heat of ablation, Z = zenithal angle of incidence, γ = coefficient of heat transfer; k is a "factor of asymmetry" in the drag.

With a high value of $k = 0.1$, and $h = 8 \times 10^{10}$ erg/g (vaporization), $\cos Z = 2/3$, $\gamma = 0.6$, $a = 6.5 \times 10^5$, we find a minimum period of rotation of

$$P_0 \geqq 0.076\, r^{\frac{1}{2}} \text{ (sec)}, \tag{4–14}$$

and an upper limit of equatorial velocity of

$$u \leqq 83\, r^{\frac{1}{2}} \text{ (cm/sec)}, \tag{4–15}$$

as a condition for the conversion of rotation into oscillation. In most cases of small meteors, $r < 10$ cm, these conditions will not be fulfilled and rotation will continue until disintegration.

Besides, major oscillations that replace rotation will also have an equalizing effect on the contour parameter. Further, a nonrotating, asymmetrical body will start oscillations when entering the atmosphere, and if there is "contour asymmetry," the drag being not equal for a rotation by 180° (like for the Robinson anemometer cups), rotation may be induced by aerodynamic pressure itself. This does not necessarily require conditions of hydrodynamic flow. The rotational velocity so acquired will be in proportion to the loss of translational velocity. For Robinson cups the proportionality factor is about 0.3–0.4. With as low a factor as 0.01, and a loss in translational velocity of 10 km/sec for a meteorite, the induced rotational velocity may be as high as 10^4 cm/sec, well above the "natural" limit.

The conclusion is that compact meteors and dust particles of all classes, except the largest crater-producing meteorites, are likely to rotate or oscillate fast enough to justify the use of $C = \frac{1}{4}$ even in individual cases, and the assumption of uniform ablation all over the surface of the meteoroid. The actual ablation of a solid meteoroid will of course depend upon the angle between the axis of rotation and the direction of motion, and will not be quite uniform.

4.3 Aerodynamic Pressure, Drag, and Energy Transfer

a. *Unshielded impact.* We consider here first the case of Problem 1, when the mean free path exceeds the radius of the meteoroid. At meteor speeds the thermal velocities of the air molecules can be neglected, and, in the absence of an air cap, each molecule hits the surface of the meteoroid and rebounds from it without interference from the others. The velocities of the impinging air molecules relative to the meteoroid can be assumed to be constant and equal to w, the velocity of the meteor. Aerodynamic pressure (normal to the surface) and drag (resistance in the direction of motion) are thus measured by the transfer of momentum in individual collisions, reckoned per unit of time and area and projected onto the corresponding directions.

Let w' be the velocity of the reflected molecule relative to the meteor, and let

$$\kappa = 1 - (w'/w)^2, \tag{4-16}$$

$$w' = qw, \qquad q = \sqrt{1 - \kappa}. \tag{4-17}$$

κ is the apparent *coefficient of accomodation*, or the fraction of the kinetic energy of air lost in the collision; most of it is transferred to the surface of the meteoroid. If the thermal agitation of the meteoroid is neglected, κ is practically equal to κ_0, the "absolute" coefficient of accommodation.

The exact relation between κ and κ_0 is given by

$$\kappa/\kappa_0 = 1 - 3kT/(\mu w^2 + 3kT_a), \tag{4-18}$$

where k = Boltzmann's constant, T = temperature of meteoroid, T_a = temperature of atmosphere, and μ = average mass of the air molecules. The difference is important only when w is small.

Consider an element of surface dS hit by the air stream at an angle of incidence α to the normal (Fig. 4). The air molecules will be reflected with some spread in angle, owing to the wavy character of the interaction potential surfaces at the solid (or liquid) boundary of dS. The *average* direction, however, can be assumed to be that of specular reflection, forming an angle 2α with the incident beam. The difference of the velocity components before and after the collision will then be

$$\Delta w_n = w \cos \alpha - (-w' \cos \alpha) = w \cos \alpha(1 + q), \tag{4-19}$$

in the direction of the normal, and

$$\Delta w_s = w - (-w' \cos 2\alpha) = w(1 + q \cos 2\alpha) \quad (4\text{–}20)$$

in the direction of incidence.

The mass of air incident per cm^2 and sec is $\rho w \cos \alpha$. Hence the local normal pressure becomes $\Delta w_n \cdot \rho w \cos \alpha$, or

$$p_n = (1 + q)\rho w^2 \cos^2 \alpha, \quad (4\text{–}21)$$

and the local drag

$$p_s = \rho w^2(1 + q \cos 2\alpha) \cos \alpha, \quad (4\text{–}22)$$

both reckoned *per unit of surface area* (not cross section); also, Eqs. 4–21 and 4–22 are valid only for $0 < \alpha < \pi/2$; for $\alpha > \pi/2$, both p_n and p_s are equal to zero, on account of shielding.

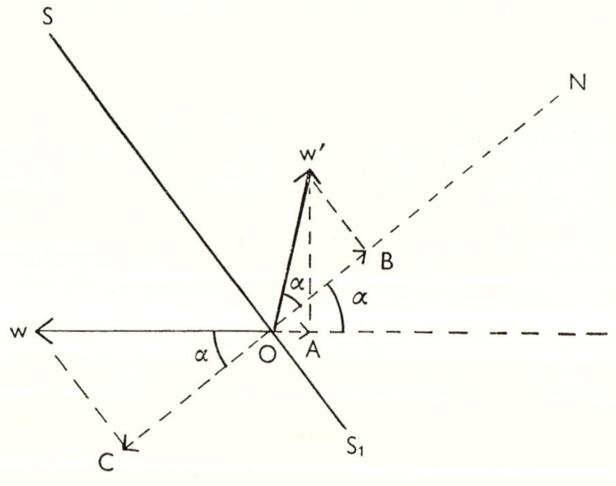

Fig. 4. Momentum transfer in atomic impact. SS_1 = surface of impact. Angle of incidence = NOA = w'ON = α. Ow = w; Ow' = w'; OC = $w \cos \alpha$; OB = $-w' \cos \alpha$; OA = $-w' \cos 2\alpha$.

The total drag depends on orientation and, for all bodies except those of spherical shape, is variable. The best we can do is to derive an average. Assuming q = const, the average value of the drag in random orientation, from integration of

Eq. 4–22 according to Eq. 4–7 (limits from 0 to $\pi/2$), becomes

$$\bar{p}_s = \tfrac{1}{4}\rho w^2, \tag{4–23}$$

and, with Eq. 4–8, the average aerodynamic resistance or the decelerating force in unshielded impact is

$$f = \tfrac{1}{4}S\rho w^2 = \bar{\sigma}\rho w^2. \tag{4–24}$$

The coefficient of accommodation has cancelled out in the process of integration.

For random orientation the average amount of energy transferred to the meteoroid per unit time is $\frac{1}{4}Sw\rho \cdot \frac{1}{2}\gamma w^2$, or

$$U = \gamma S\rho w^3/8 = \tfrac{1}{2}\gamma\bar{\sigma}\rho w^3. \tag{4–25}$$

Here γ = coefficient of heat transfer; for unshielded impact $\gamma \leqq \kappa$; γ differs from κ by the amount spent on impact radiation, dissociation, and excitation of the reflected air molecules, and sputtering. The difference is presumably small.

With Eq. 4–3, Eq. 4–25 transforms into

$$U = \pi B\gamma r^2\rho w^3/6. \tag{4–26}$$

The work of aerodynamic friction per unit time is $U_f = fw$, or according to Eq. 4–24 as in Problem 1,

$$U_f = \tfrac{1}{4}S\rho w^3 = \bar{\sigma}\rho w^3. \tag{4–27}$$

Here the system of reference is air at rest. The difference $(U_f - U)$ is chiefly accounted for by the kinetic energy carried away by the reflected air molecules; as $\gamma < 1$, the difference is always greater than $\frac{1}{2}U_f$.

In the preceding formulae, the thermal velocities of the air molecules have been disregarded. For small meteor velocities the rather complex conditions can be excellently represented by the following approximations. Denoting by u the quadratic mean thermal velocity component in one coordinate, $u^2 = kT/\mu$, the average drag for $w \gg u$ can be set equal to

$$\bar{p}_s = \tfrac{1}{4}\rho(w^2 + 1.5u^2), \tag{4–28}$$

$$f = \bar{\sigma}\rho(w^2 + 1.5u^2). \tag{4–29}$$

For $w \leqq u$, the formulae for drag are

$$\bar{p}_s = 0.625\rho uw, \tag{4–30}$$

$$f = 2.5\rho\bar{\sigma}uw, \tag{4–31}$$

valid, of course, only when the mean free path is greater than the radius.

b. *Drag in shielded impact.* When there is a hydrodynamic air cushion, i.e., when $\lambda < r$, where $\lambda =$ half-energy range *at the velocity of the meteor*, the aerodynamic pressure and drag upon an element of surface are influenced by the shape of the entire body and cannot be simply calculated. The average drag force for random orientation can be generally represented as

$$f = K\bar{\sigma}\rho w^2, \tag{4–32}$$

where $\bar{\sigma} = \frac{1}{4}S$, and K is the coefficient of drag. Normally $\frac{1}{4} < K < 1$. Table IV suggests values of K for high supersonic velocities as used by Öpik in a theory of meteorite ablation (to be published).

TABLE IV

Kinetic Thickness and Drag Coefficient

$d = 0.75r/\lambda$	K
0	1
1	0.75
2	0.625
4	0.562
∞	0.5

Here d is the average kinetic thickness of the air cap, in units of the half-energy range.

Table IV represents the assumed transition of drag from Problem 1 ($d = 0$) to Problem 2 ($d \to \infty$). The value $K = 0.5$ is based on a calculated value for an actual hydrodynamic model of flow. The table can be used in the case of random orientation of a model of not too unusual shape.

For less than sonic velocities and $d \to \infty$ (or large), $K = \frac{1}{4}$ can be adopted. For a flat surface moving in the direction of its normal the values of K are to be doubled.

For low velocities the viscosity of the medium becomes important. Stokes' formula as corrected by Millikan from experiments with oil droplets in air, and valid only for $u > w$, i.e., when the velocity of drift is smaller than the molecular velocity,

$$f = 6\bar{\sigma}\eta w/(r\psi), \qquad (4\text{–}33)$$

should be used, where η is the coefficient of viscosity and

$$\psi = 1 + \frac{\lambda_0}{r}\,(0.864 + 0.290e^{-5r/4\lambda_0}). \qquad (4\text{–}34)$$

In the earth's atmosphere approximately

$$f = 1.02 \times 10^{-3}\,\bar{\sigma}w/(r\psi). \qquad (4\text{–}35)$$

The factor ψ extends the formula into the transition region $\lambda_0 > r$ and beyond. Of the rival formulae 4–32 and 4–33 or 4–35, that yielding the greater drag must be used. For small velocities and radii this will be formula 4–33, for large velocities, formula 4–32.

For large λ_0/r ratios Eq. 4–33 transforms into $f = 2.76\bar{\sigma}\rho u w$, to be compared with Eq. 4–31. This is obtained by substituting $\eta = \frac{1}{3}\lambda_0\,\rho\bar{v}$, $\bar{v} = (8/\pi)^{\frac{1}{2}}\,u$ into Eq. 4–33, and $\psi \to 1.154\lambda_0/r$ for $\lambda_0/r \to \infty$.

c. *Energy transfer in shielded impact.* In Problem 1, the mass of vapors formed may considerably exceed the mass of intercepted air. A vapor layer may arise, preventing direct impact of the air molecules on the meteoroid. The heat transfer coefficient, γ, may differ considerably from κ. The immediate heating is restricted to the top of the vapor layer; the energy reaching the meteoroid is carried by fast penetrating molecules, either direct or created in secondary collisions. The problem is simlar to that of multiple scattering of light in a diffuse medium. Calculations applied to a certain model[6] yielded values of the heat transfer coefficient as contained in Table V. In this table $d_v = \Delta r/\lambda$ is the kinetic thickness of the vapor layer; Δr = linear thickness, λ = half-energy range in the vapor layer, at about one-half the velocity of the meteor.

TABLE V

Heat Transfer Through Vapor Layer
(without air cap)

d_v	γ_v
0	κ
1	$1/2(\kappa + 0.58)$
2	0.58
3	0.43
4	0.32
> 4	$1.28/d_v$

In Problem 2 the shielding is by the air cap which transmits its energy to the meteoroid through contact. The type of flow is here relevant for the heat transfer. If v is the velocity of flow and λ the half-energy range in the air cap, the viscosity in the air cap is of the order of $\eta = \eta_0 v\lambda/(u\lambda_0)$, where η_0 = viscosity, u = molecular velocity, and λ_0 = half-energy range in the atmosphere. Substituting this into Eq. 2–3, with $\eta_0 = 1.7 \times 10^{-4}$, $u = 4.8 \times 10^4$, $\lambda/\lambda_0 = 10$ (from 8 to 14 at meteor velocities, cf. Section 6.1), and substituting $0.75\ \rho r$ for the product ρL (Section 6.1), we obtain for the Reynolds number in the air cap (with ρ = density in the free atmosphere)

$$Re = 2.1 \times 10^7 \rho r. \tag{4–36}$$

For laminar flow $Re < 1000$, or

$$r = 4.8 \times 10^{-5}/\rho. \tag{4–37}$$

According to this, the upper limit of radius for laminar flow is as follows:

Altitude, km	40	50	60	70
r, cm	< 12	43	125	500

Thus, except for meteorites which penetrate deep into the atmosphere, laminar flow is generally valid. The air cap has the properties of a sticky fluid, and the transfer of heat is chiefly by conductivity.

Conductivity depends on the kinetic properties of the gas. It is possible to express the heat transfer directly in terms of the kinetic thickness, d, of the air cap.

For an extended air cap the case of heat transfer from an infinite wall can be applied, in the simplified setting of Section 6.7. The conductivity is given by

$$k_t = (\tfrac{2}{3})\lambda_c \rho_c v_c c_v\,, \tag{A}$$

the factors to the right denoting half-energy range, density, mean molecular velocity, and specific heat in the air cap. The flow of heat toward the meteoroid is

$$Q = -\,y_0' \cdot k_t = \tfrac{1}{2}\gamma \rho w^3, \tag{B}$$

where y_0' = temperature gradient at the surface.

With 0.75 of the kinetic energy of the air molecules available for heating the air cap (the rest appears as kinetic energy of flow), the asymptotic temperature of the air cap, i.e., outside the cooling influence of the meteoroid, is

$$T = \tfrac{3}{8} w^2/c_v. \tag{C}$$

With the "parabolic" simplification of the heat-transfer problem and $k = \frac{2}{3}$ in Eq. 6–27, it can be shown that

$$-\,y_0' = T/(2\gamma x), \tag{D}$$

where x is the linear thickness of the air cap. Further, in the "affected" portion adjacent to the meteoroid, $T_c \cong \frac{2}{3}T$, $v_c \cong w/\sqrt{2}$. By definition,

$$\lambda_c = x/d, \tag{E}$$

d being the kinetic thickness. Hence Eq. B, after substitution of Eqs. A, C, and E, becomes

$$\gamma = (\rho_c/\rho)(4\sqrt{2}\,\gamma d)^{-1}. \tag{F}$$

Further, $\rho_c/\rho = p_c T_0/pT_c$, where p and T_0 are pressure and temperature in the free atmosphere, p_c and $T_c = T(1-\gamma)$—in the air cap; $p = \frac{1}{3}\rho v_2^2$, $p_c = \frac{1}{2}\rho w^2$ (v_2 = gas-kinetic velocity in the atmosphere). Thus $p_c/p = 1.5w^2/v_2^2$, $T_0/T_c = 4v_2^2/[3w^2(1-\gamma)]$, and

$$\rho_c/\rho = 2/(1-\gamma). \tag{4–38}$$

Substituting this into Eq. F, a nondimensional formula is obtained, valid for $d > 8$:

$$\gamma^2(1-\gamma) = (2\sqrt{2}\,d)^{-1}. \tag{4–39}$$

Similarly, for a thin cap with $d < 8$,

$$\gamma(1-\gamma)^{\frac{1}{2}}/(0.75-\gamma) = 2\sqrt{3}/d. \tag{4–40}$$

Expression 4–40 is valid only when it leads to values of γ greater than those of Table V, or for $d > 2.5$. For $d < 2.5$ the values of Table V are to be taken, as these represent minimum values of γ corresponding to immediate penetration of the kinetic energy flux.

TABLE VI

Heat Transfer Through Air Cap

d_c	γ_c
0	κ
1	$0.5(\kappa + 0.58)$
2	0.58
3	0.45
4	0.39
6	0.30
8	0.25
11	0.20
18	0.15
40	0.10
150	0.05
> 150	$0.6d^{-\frac{1}{2}}$

Hence, from Eqs. 4–40, 4–39, and with the imposed restriction, the values of heat transfer as contained in Table VI are suggested.

Here $d_c = 0.75r/\lambda$ is the kinetic thickness of the air cap, λ being the half-energy range in the free atmosphere at about one-half the velocity of the meteor. At very large values of d_c, of the order of 10^4 and greater, radiation from the air cap becomes an important source of energy transfer. This, however, leads us into the problem of meteorite flight which is too complicated to be tackled within the framework of this tract; the condition of laminar flow will also no longer be fulfilled in such a case.

We note that over the range of small d-values Tables V and VI are practically identical, as well as their asymptotic expressions. For order-of-magnitude estimates, and within the critical range of an incipient air cap, $d = 2$–4, it is permissible to use the expression

$$\gamma \cong 1.28/d, \tag{4–41}$$

equally for the air cap and the vapor layer.

When there are simultaneously present an air cap of kinetic thickness d_c, and a vapor layer d_v underneath, the resulting heat transfer coefficient can be set equal to

$$\gamma = \gamma_c \gamma_v. \tag{4–42}$$

4.4 Shielding by Vapors

Even before an air cap is formed, the vapors from the meteoroid may build up a blanket in front of the meteoroid. The ratio of the mass of vapors (ξ) to that of the incident air stream (A_s) is

$$\xi/A_s = \tfrac{1}{2}\gamma w^2/h, \tag{4–43}$$

where h is the heat of vaporization. With $h = 6 \times 10^{10}$ erg/g (evaporation from a previously molten meteoroid) and $\gamma = \kappa = 0.6$, this yields

$$\xi/A_s = 5 \times 10^{-12} w^2. \tag{4–44}$$

For the lowest velocity, $w = 1.2 \times 10^6$, the ratio is 7.2. Despite this, the blanketing role of vapors is relatively small. One reason for this is that the distance to which a vapor molecule can recede from the surface of the meteoroid in a forward direction, of the order of

$$\Delta = \lambda' u/w \cong 10^{-2}(\lambda\lambda_0)^{\frac{1}{2}}, \tag{4-45}$$

or the linear depth of the vapor coma, is small (Fig. 5). Here $\lambda' = (\lambda\lambda_0)^{\frac{1}{2}}$ is the effective free length of path, and $u \cong 5 \times 10^4$ the gaskinetic velocity of vapor molecules (at $T \sim 2100°$ K). Only a thin vapor layer can be built up in front of the meteoroid. Another reason is the fact that the

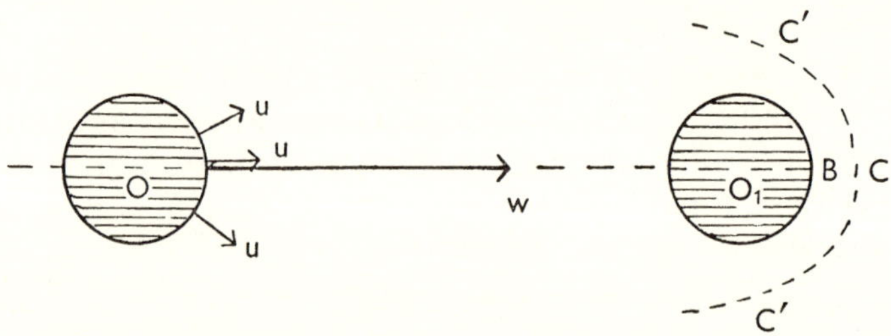

Fig. 5. Formation of vapor layer in front of the meteoroid. At O, vapor molecules depart from the surface with small relative velocities u, while the meteoroid moves on with a large velocity w. $OO_1 = \lambda'$ is the path over which the vapor molecule is turned back by a collision with an air molecule. $BC = \Delta$ is the thickness of the vapor layer, and CC' the coma front. $\Delta/\lambda' \cong u/w$.

blanketing effect of the layer reduces the heat flow and evaporation and works thus against a further increase in the thickness of the layer. Allowing for the dependence of λ upon velocity, and for the radial flow of vapors as restricted by Eq. 4–45, an expression for the kinetic depth of the vapor layer at normal incidence has been obtained:

$$d_v = \gamma w^2\{32h[1 + \lambda' u/(rw)]\}^{-1}. \tag{4-46}$$

The kinetic depth for oblique incidence is $d_v \sec \alpha$ when $\Delta \ll r$, and d_v for $\Delta \gg r$. For $r \to \infty$, $d_v = d_{\max} = \gamma w^2/(32h)$.

For stony meteors, with their low conductivity of heat, melting and vaporization normally takes place simultaneously, or "on the spot," and the effective value of h can be set equal to 8×10^{10}; with this and Eq. 4–41 when $d > 2$ (but $\gamma = 0.6$ when $d < 2$), we find

$$d_v = 7.1 \times 10^{-7} w[1 + \lambda' u/(rw)]^{-\frac{1}{2}}. \tag{4-47}$$

For oblique incidence and $\Delta \ll r$ the divisor shall contain the additional factor $(\cos \alpha)^{\frac{1}{2}}$.

Table VII contains the values of $[(d(\cos \alpha)^{\frac{1}{2}}]_{max} = 7.1 \times 10^{-7} w$ and γ_{min}, corresponding to $r \to \infty$ ($r > 0.1$ cm).

TABLE VII

Velocity and Lower Limit of Heat Transfer in Vapor Shielding Without Air Cap; Normal Incidence

w, km/sec	d_{max}	γ_{min}
≤ 29.4	≤ 2	($\kappa = 0.60$)
42.2	3	0.43
56.2	4	0.32
84.3	6	0.213

Thus, for low and moderate velocities (below 40 km/sec) the shielding effect of the vapors is negligible.

The effect decreases with decreasing size and is negligible for micrometeors.

Table VII and Eq. 4–47 exaggerate the screening capacity of vapors by assuming their free escape. Actually, the vapors are suppressed by the air stream. Even in the absence of an air cap, when a vapor layer of $d \simeq 2$ has accumulated and when $\gamma \cong \kappa$ still holds, the layer will absorb a considerable fraction of the aerodynamic momentum, and aerodynamic pressure will counteract the forward flow of vapor. The writer has considered this effect for the case of surface equalization of evaporation (e.g., by rotation).[6] It turns out that, when the vapor pressure is uniform all over the surface, near the frontal apex of the meteoroid and down to an angle of incidence α_0 the aerodynamic pressure exceeds the vapor pressure and a frontal "bald cap" from $\alpha = 0$ to $\alpha = \alpha_0$ is formed, with

$$\cos \alpha_0 = \kappa/(\kappa + b),$$

$$\bar{\gamma} = \kappa b(2\kappa + b)/(\kappa + b)^2, \tag{4–48}$$

and $b = 4 \times 10^6/w$. Eq. 4–48 gives the resulting total heat transfer in random orientation and for $r > 0.1$ cm. With $\kappa = 0.6$ the values of Table VIII result.

TABLE VIII
Shielding Vapors for Constant Vapor Pressure All Over the Surface; $r \geqq 0.1$ cm

w, km/sec	α_0	$\bar{\gamma}/\kappa$
16	78.8°	0.963
25	74.2	0.926
40	68.0	0.859
60	61.4	0.785
80	57.0	0.703

For smaller radii the values of $\bar{\gamma}/\kappa$ are increasing and approach 1 at $r \to 0.04$ cm, being half way in between at $r \cong 0.06$ cm. Thus, in the case of all-sided evaporation, the heat transfer is very little affected by the vapor layer. In the absence of an air cap, the only important case of shielding by vapors is that of Table VII, for $w > 40$ km/sec and vaporization "on the spot" from the front side only.

4.5 The Coefficient of Accommodation

The energy lost in collision can be transformed into 1, thermal energy, 2, impact radiation, 3, energy of sputtering, and 4, excitation or ionization, chiefly of the reflected atom. The first component takes the major share and, in calculating the heating of the meteoroid, the other three can be neglected. The thermal energy results from the vibration of the atoms of the lattice, as well as from radiationless conversion of excited quantum states induced in the solid by the collision.

The transfer of energy greatly depends upon the mass ratio and the kinetic energy of impact. The equipotential surfaces of atomic interaction (mostly repulsive) are smooth at first at low values of the energy and at a great distance from the lattice level; they become more undulatory when the level is approached with greater energies and the field of individual atoms is felt; at still higher energies (exceeding the lattice energy), the surfaces become closed around each atom, and the potential energy steeply increases as the atomic nuclei are approached (Fig. 6). Therefore, impact at low energies (e.g., at thermal velocities) is countered by the

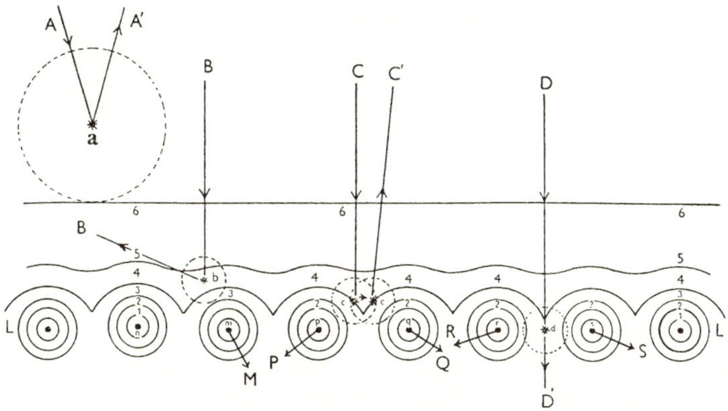

Fig. 6. Atomic impact into solid (or liquid) surface. The black dots on the line LL represent surface lattice atoms. Concentric circles 1, 2, 3 (atom n) are equipotential surfaces in the order of decreasing energies; surfaces 4, 5 are merged, and 6 is a smooth common surface corresponding to thermal energies. The impinging atoms are shown as asterisks, with broken circles indicating their effective radii corresponding to their kinetic energies. Atom a, of thermal energy and, thus, of large radius, impinges from the direction Aa and is reflected by surface 6 toward aA′. Atom b, of high (meteoric) velocity and, accordingly, of small radius, impinges from Bb, penetrates until contact with surface 3 of m, whence it is reflected back into space toward bB′. The potential energies decrease outwards exponentially; therefore potential surfaces of lower grade, 4, 5, and 6, have little effect on the motion of b. The lattice atom m receives a vibrational impulse mM in the collision. Another fast atom c impinges from Cc, is reflected by surface 2 of p toward c′, thence in a collision with surface 2 of q it is diverted back toward c′C′. The two lattice atoms hit receive vibrational impulses pP and qQ. Atom d, coming from dD, passes the "hole" between surfaces 2 of r and s and proceeds further toward dD′, being ultimately trapped in the lattice (penetration trap).

combined smooth potential of repulsion of several cooperating atoms; because of the greater combined mass, they will take up less vibrational energy (the energy transfer decreases with the mass ratio), and the impinging atom will be reflected with less relative loss of kinetic energy than in the case of high-energy impact which can be counteracted only by the "private" electron shell of an individual atom. This is the main cause of the increase of κ_0 with temperature at thermal velocities.

An approximate evaluation of κ can be made by disregarding quantum transitions and considering only the inertial elastic interaction of the colliding systems.

At meteor velocities the kinetic energy greatly exceeds the lattice energy, and the collision is taking place practically atom with atom. An atom of the lattice either alone expels the intruder or reflects it towards another member of the lattice; the vibration acquired in the collision it transmits to its neighbours (Fig. 6).

Taking the atom in the lattice as system of reference (and neglecting the relative motions in the lattice), let μ be the mass, w the velocity of the impinging atom, and μ_1 the mass of the atom in the lattice. The mass ratio is $m = \mu/\mu_1$, and the kinetic energy $E = \frac{1}{2}\mu w^2$.

According to the laws of conservation of momentum and energy, in a central-force model of short-range interaction,[7] the potential energy of interaction (repulsion) at periastron is

$$V_0 = E_r \cos^2 i, \tag{4–49}$$

where i is the periastron angle, or the angle between the line joining the two atoms at closest approach, and the original direction of incidence; the relative kinetic energy is

$$E_r = E/(1+m). \tag{4–50}$$

For a completely elastic collision the transfer of kinetic energy from μ to μ_1 is

$$\Delta E = 4V_0 m/(1+m). \tag{4–51}$$

The coefficient of accommodation is then

$$\kappa = \int_b^c (\Delta E/E)\mathrm{f}(i)\,di \Big/ \int_b^c \mathrm{f}(i)\,di, \tag{4–52}$$

where $f(i)$ is the probability of the angle to be between i and $i + di$; b and c are specifically defined limits of integration.

Assuming further $f(i) = 2\pi \sin i \cos i$ as for an elastic sphere, with Eqs. 4–51, 4–49, and 4–50 the integral in Eq. 4–52 becomes

$$\kappa = 2m(\cos^2 b + \cos^2 c)/(1 + m)^2. \tag{4–53}$$

The relative area of the cross section is $\cos^2 b - \cos^2 c$. The limits of integration in a two-atom collision are from $b = 0$ to $c = \pi/2$. In the case of a lattice, $c < \pi/2$ because penetration is limited by other lattice atoms; the impinging atom is reflected inwards, meets another member of the lattice and loses more energy in a second collision before being finally reflected (Fig. 6); also, for oblique incidence small values of i are screened by neighboring atoms of the lattice.

Let θ be the direction of motion of the reflected atom after the primary collision, reckoned relative to the original direction,

$$\tan \theta = \sin 2i/(m - \cos 2i). \tag{4–54}$$

At normal incidence to the lattice surface ($\alpha = 0°$) the upper limit $c = i_1$ for a single collision to take place corresponds to $\theta = \pi/2$, or

$$\cos 2i_1 = m. \tag{4–55}$$

The calculation of κ at normal incidence is thus to be split up into two stages: a component of κ obtained by integration from $i = 0$ to $i = i_1$, corresponding to a single collision; and another from $i = i_1$ to $i = i_2$ which undergoes a second collision. $i_2 < \pi/2$ depends upon how deeply the intruder can squeeze itself between the atoms of the lattice; it increases with velocity and for meteors $i_2 = 60°$ can be assumed. In the second collision, the limits can be assumed as from 0 to i_1 only. When $m > 1$, $\tan \theta > 0$ and the atom is always reflected inwards, the collisions being thus at least double. Table IX contains the results of the calculation.

At oblique incidence the coefficient of accommodation will be smaller because of shielding of the lattice surface by other atoms. The degree of shielding depends on the effective value of the atomic radius corresponding to potential energy of interaction $V \cong E_r$. At meteor velocities it is estimated that the limit of θ for being reflected back after one single collision at 60° incidence is from 30° to 90°, with an average of $\bar{\theta} \cong 50°$; hence

$$\sin 2i_1/(m - \cos 2i_1) = +1.20 \quad (60° \text{ incidence}),$$

and $i_2 = 90°$. The assumptions made imply a minimum distance in a central collision equal to one-half the lattice spacing or about 1.25×10^{-8} cm (Fig. 6). At this distance the energy of interaction (N → Fe) is $V = E_r = 80$ ev[7], or $E \cong 100$ ev, corresponding to a velocity of about 35 km/sec. In the 5th line, κ' is the energy transfer coefficient in a collision of two free atoms. This is essentially the quantity assumed by Sparrow[4] who quotes $\kappa = 0.27$ to 0.37 for impact by nitrogen.

Assuming the dependence of κ upon the angle of incidence to be of the form

$$\kappa/\kappa_n = (\cos \alpha)^n, \tag{4–56}$$

the values for $\alpha = 0°$ and $\alpha = 60°$ lead to values of the exponent n as given in the 9th line of Table IX.

TABLE IX

Accommodation Coefficients Calculated with the Elastic Collision Model (for $m = 0$, $\kappa = 0$)

Lattice atom	O	Mg-Si	Fe	Stone average
Atomic weight, A_t	16	26	56	23
% atoms in stone	57	36	7	100
$m = 14.5/A_t$	0.906	0.558	0.259	0.63
$\kappa' = 2m/(1+m)^2$	0.498	0.459	0.327	—
i_1 (Eq. 4–55); 1st collision	3.9°	28.0°	37.5°	—
κ_n, 0° incidence	0.989	0.876	0.600	0.921
κ, 60° incidence	0.799	0.651	0.428	0.720
n	0.31	0.43	0.49	0.36
Average κ (0°–90°)	0.855	0.722	0.482	0.781

It can be shown that in random orientation, or for a sphere, and with Eq. 4–56, the average value of the coefficient of accommodation is

$$\bar{\kappa} = 2\kappa_n/(n+2). \tag{4–57}$$

This is given in the last line of Table IX.

The mass ratios were here taken as for single atoms (N or O), and not as for diatomic molecules; this is because the kinetic energy greatly exceeds that of the chemical bonds, so that interaction takes place with the single components of a molecule.

Van Voorhis and Compton[39] determined accommodation coefficients of ions of inert gases (A^+, Ne^+, He^+) at impact on metal spheres (Mo, Pt) of about 3 mm diameter. The ions were attracted by the negative charge of the sphere, striking its surface in a radial direction with energies of 21–141 ev. The values thus correspond to normal incidence, $\alpha = 0°$. A graphical comparison of κ_n observed[39] with that calculated (Table IX) shows that both are smooth functions of the mass ratio, but there is some systematical difference; for $m < 0.4$ the observed values are higher than those calculated. Apparently, quite a considerable fraction of the energy is spent on quantal transitions in the lattice, and the accommodation coefficient at small mass ratios is thus increased. The combined observed—calculated curve can be regarded as representing the probable dependence of the accommodation coefficient upon mass ratio; the values of κ as read from the curve are given in Table X.

TABLE X

Probable Values of the Accommodation Coefficient ($E \cong 100$ ev)

Impinging atom	—	N; O	N; O	N; O	N; O
Lattice atom	—	Fe	Mg; Si	O	Stone ave.
m	0	0.259	0.558	0.906	—
κ_n	0.450	0.650	0.876	0.989	0.924
n	(1.)	0.49	0.43	0.31	—
Spherical $\bar{\kappa}$	0.300	0.522	0.722	0.855	0.785

At higher velocities penetration into the lattice may take place; an atom that has penetrated through the first lattice layer is effectively trapped and $\kappa = 1$ practically results (Fig. 6). The conditions of penetration have been studied for an iron lattice; the results are contained in Table XI. On account of penetration, the accommodation coefficient increases somewhat,

$$\kappa = y + (1 - y)\kappa_1,$$

κ_1 being the value of the coefficient without penetration. The increase is relatively small, contrary to what has been formerly suggested.[6]

TABLE XI

Penetration into Lattice

Velocity of N atom, km/sec	42	56	67	89	107
Probability of penetration, y	0.01	0.04	0.09	0.16	0.25

Roughness of the surface may considerably affect the effective value of the coefficient of accommodation in unshielded impact. The values of κ considered above refer to smooth surfaces which can be described by the term "polished," as well as to liquids. Cavities cause multiple reflections of the air molecules and increase κ. The effect is difficult to allow for in a general way, as it depends upon the specific nature of roughness. It is important only in the theory of solid micrometeors which presumably have pitted surfaces, the result of bombardment by other particles in space. Their surface must be covered with pits whose frequency increases with decreasing size according to a law similar to that of meteor sizes. The following model may be a realistic approach to these conditions (Fig. 7). Let the pit profiles be similar to carborundum pits on glass, and let the surface be covered to 75 per cent by large crater bowls "of the 1st order." The interior of each bowl, as well as the interstices between the first order craters let be covered in a similar manner by smaller bowls of the 2nd order, their diameters being equal to 1/3 of those of the 1st order. These in their turn let be covered by 3rd order cavities, with linear dimensions three times smaller than those of the 2nd order, and so on. A geometrical progression of cavities decreasing in size, or a "geometric turbulence" results. Let the average

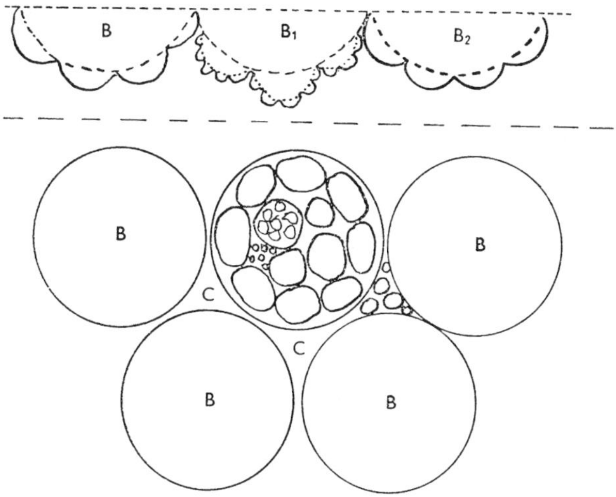

Fig. 7. A model of rough surface. B = first order crater bowl; C = interstice. The upper portion is a vertical, the lower portion is a horizontal cross section. B, B_2—1st and 2nd order cavities only; B_1—1st, 2nd, and 3rd order cavities.

outlet of a bowl, or the solid angle of the opening as viewed from the bottom be 0.5 of a hemisphere. If κ_i is the coefficient of accommodation of the ith order, that of the $(i - 1)$st order, after taking into account multiple reflections becomes $\kappa_{i-1} = 2\kappa_i/(1 + \kappa_i)$ for the bowl, and $\kappa_{i-1} = \kappa_i$ for the "flat" interstices. Weighting these two values in the ratio of 3 to 1, we obtain the average

$$\kappa_{i-1} = \tfrac{1}{4}\kappa_i(7 + \kappa_i)/(1 + \kappa_i). \qquad (4\text{–}58)$$

The highest (ith) order of roughness is that at which κ has been determined experimentally, i.e., that corresponding to a "polished" surface. Optical surfaces are polished to about 2.5×10^{-6} cm. We assume the ith order of geometric turbulence to consist of elements of $D_i = 2 \times 10^{-6}$ cm diameter, and the first order to have a diameter equal to the average thickness of the meteoroid, $D_1 = r/B$ (Eq. 4–5).

Setting $B = 5$ as an average, we obtain $D_1 = \frac{1}{5}r$. The number of orders (in a progression of 3 to 1) is then

$$i = 1 + \log (D_1/D_i)/\log 3 = 11 + 2 \log r \geqq 1. \quad (4\text{–}59)$$

According to Eq. 4–58, with $\kappa = 0.6$ for $i = 1$, the accommodation coefficient as depending on the number of orders of roughness is given in Table XII.

TABLE XII

Accommodation Coefficient and Roughness of Surface

i	1	2	3	4	5	6	7	∞
κ_i	0.600	0.713	0.802	0.867	0.913	0.944	0.964	1.000

Table XIII gives the number of orders of roughness according to Eq. 4–59.

TABLE XIII

Radius and Number of Orders of Roughness

r, cm	1×10^{-4}	1×10^{-3}	2×10^{-3}	4×10^{-3}	8×10^{-3}
i	3.0	5.0	5.6	6.2	6.8

Allowance for roughness has been made by the writer in the theory of micrometeors (to be published); for ordinary meteors there is no need of it because of shielding effects. For iron micrometeors an interpolation formula, closely representing the figures of Tables XII and XIII and valid for $r < 0.04$ cm, proved useful:

$$\log \kappa = 0.035 + 0.025 \log r. \quad (4\text{–}60)$$

4.6 Radiation from Meteoroids

Radiation from the surface of the meteoroid is important in micrometeors and in the early stages of ordinary meteors. It is given by Stefan's formula

$$Q = \varepsilon s T^4, \tag{4–61}$$

where $s = 5.67 \times 10^{-5}$ (erg cm^{-2}sec^{-1}deg^{-4}), and $\varepsilon = \varepsilon(T)$ is the average emissivity at the given temperature. For a black body $\varepsilon = 1$.

The original rough surfaces may approach black-body conditions; molten smooth surfaces may deviate considerably. For stone the surface during the flight through the atmosphere may remain nearly a black body, whereas molten iron will not. The reflectivity of metals increases with wavelength, and their emissivity is rather low at temperatures below 3000° K which alone are here of importance.

In calculations of iron micrometeors, the writer assumed the emissivity of the smooth solid or liquid to be that of "steel, untempered."[40] Averaged over Planck's curve, the emissivity is well represented by $\bar{\varepsilon} = 0.193(T/1500)^{0.65}$, and the adopted radiation formula for a smooth iron surface becomes

$$Q = 9.64 \times 10^{-8}\, T^{4.65}, \tag{4–62}$$

valid for the temperature interval 1500–3000° K.

For a rough surface, of a model described in Section 4.5, the emissivity increases from multiple reflections according to formula 4–58, with ε substituted for κ. The lower limit of roughness is estimated at half-wavelength, the upper limit $D_1 = \frac{1}{5}r$ as before; hence the number of orders of roughness becomes

$$i_\varepsilon = 0.2 + 2 \log (r/L) \geqq 1, \tag{4–63}$$

where L is the effective wavelength. For the relevant interval of temperature the radiation formula for a rough iron surface and the chosen model of pitting becomes

$$Q = 9.53 \times 10^{-8}\, T^{4.65}(1000\, r)^{0.3}, \tag{4–64}$$

valid for $(0.1) > r > 1.040 \times 10^{-3}$ cm; below $r = 1.040 \times 10^{-3}$ formula (4–62) as for a smooth surface shall be used.

4.7 Sputtering

The impact of high-speed atomic particles may knock out single atoms from a solid or liquid surface, a phenomenon which is called sputtering; unlike vaporization it does not much depend upon temperature. There does not yet seem to be a satisfactory theory of it.[41]

The impinging atom leaves a fraction κ of its kinetic energy in the lattice; the vibrations of the lattice propagate in all directions and, before the energy is dissipated over the volume of the solid, a few surface atoms near the point of impact may acquire vibrational energies in excess of the lattice energy and become thus ejected. However, on a model of purely elastic propagation of the vibrations, the calculated efficiency of sputtering is much greater than that observed.

A very satisfactory representation of the experimental data has been obtained by the writer upon a model, the essence of which consists in the assumption that 0.66 of the kinetic energy is lost into inelastic (quantal) transitions inside the lattice for each collision. The agreement between calculation and experiment is of such a nature that the sputtering rates can be more or less reliably predicted. The number ν of atoms sputtered for each impinging atom can be represented as

$$\nu = 5.3 \times 10^{-3} \kappa_n (E - 32i), \qquad (4\text{–}65)$$

where E is the kinetic energy of the impinging atom, i the lattice energy, both expressed in electron volts; κ_n is to be interpolated from Table X for the appropriate mass ratio, m. [It can be assumed that $\kappa(m) = \kappa(1/m)$.]

The sputtering rates at 500 ev for 23 combinations of atoms[41] agree within a close order of magnitude with those calculated from Eq. 4–65. The differences are not greater than those between different experimental sets. The variation of sputtering for $E = 200\text{–}1500$ ev is also satisfactorily represented. It appears that Eq. 4–65 can be used with confidence also in the case of meteors. For an iron surface the mass sputtered becomes

$$\Theta = 0.0131(E - 135), \text{ (Fe/Air)} \qquad (4\text{–}66)$$

where Θ = ratio of sputtered mass to incident air mass, and E = kinetic energy of air (N or O) *atoms* in electron volts. For stone

$$\Theta = 0.0078(E - 200) \text{ (Stone/Air);} \qquad (4\text{–}67)$$

it is here assumed that the sputtered substance is atomized, and that the energy of releasing one atom is 6 ev.

From Eqs. 4–66 and 4–67 it follows that sputtering begins at $w = 42$ km/sec for iron, and 51 km/sec for stone.

4.8 Deceleration

Let f_a be the drag of aerodynamic origin, f the total drag including that caused by the meteor vapors, so that $f = \zeta f_a$. ζ can be called the *jet correction factor*. The deceleration is then $dw/dt = -\zeta f_a/M$. With Eqs. 4–32 and 4–4 this becomes

$$\frac{dw}{dt} = -\zeta K B \rho w^2/(4r\delta). \qquad (4\text{–}68)$$

ζK may be called the deceleration factor.

The jet correction depends upon vapor pressure, p_v, which can be determined from Eq. 3–2 (with $m = 50$ as for stone) when ξ, the intensity of evaporation, is known. This depends upon the heat uptake and, for a thin air cap ($d < 2$), can be set equal to

$$\xi = \tfrac{1}{2}\gamma \rho w^3 \cos \alpha / h. \qquad (4\text{–}69)$$

When $p_v < p_n$, the vapors are suppressed by aerodynamic pressure and the jet correction is absent, $\zeta = 1$. When $p_v > p_n$, the jet correction is

$$\zeta = \left(\int_0^{\pi/2} p_v \cos \alpha \sin \alpha \, d\alpha \right) \Big/ (K \rho w^2). \qquad (4\text{–}70)$$

Along these lines, from an approximate theory the deceleration factors as given in Table XIV have been calculated. The table covers stony meteors only, for which nonisothermal vaporization "on the spot," with $h = 8.1 \times 10^{10}$ erg/g, can be assumed; their radii should exceed 0.2 cm. The theory is good for $d_0 = 0$ (small meteors, no air cap) and $d_0 \to \infty$ ($\zeta K = 0.5$), whereas the transition figures are less reliable. Whatever the uncertainty in the intermediate cases, the table is sufficient to prove the wide range of variation of the deceleration factor.

TABLE XIV

Drag Coefficient with Jet Correction

ζK = deceleration factor; d = kinetic depth of the air cap. (Stony meteors, nonisothermal, vaporization "on the spot.")

w, km/sec d	12	20	30 ζK	45	60	75
0	1.00	1.28	1.92	2.88	3.84	4.80
1	0.75	1.28	1.92	2.88	3.84	4.80
2	0.69	1.28	1.92	2.88	3.84	4.80
4	0.56	0.64	0.96	1.44	1.92	2.40
6	0.54	0.54	0.64	0.96	1.28	1.60
10	0.52	0.52	0.52	0.58	0.77	0.96

The values of ζK of Table XIV more or less cover the range of the photographic material which Whipple *et al.*[10] used for the determination of atmospheric densities. He used a formula of the type of Eq. 4–68; in this, r is calculated from the observed luminosity with a certain assumed luminous efficiency, β, $r \sim \beta^{-\frac{1}{3}}\delta^{-\frac{1}{3}}$. Hence, apart from observed data, the *calculated* atmospheric density will be proportional to the following combination of assumed factors:

$$\rho \sim \delta^{\frac{2}{3}}/(\zeta K B \beta^{\frac{1}{3}}) = G. \tag{4–71}$$

Because of natural selection, it is expected that less dense meteors and those of greater relative area and velocity are observed at greater altitudes; also at greater altitudes (smaller d) and velocities ζK is greater (Table XIV), whereas the influence of $\beta^{\frac{1}{3}}$ is small. The net effect is a steady decrease of G with altitude; therefore, the atmospheric densities calculated from decelerations with an assumed constant value of G will be progressively overestimated. Whipple and his co-workers assumed indeed a constant value of G which satisfactorily represented the average atmospheric densities for the range 60–75 km,[10] as given by rocket ascents. As expected, above 75 km systematically greater densities resulted from the deceleration method, as compared with the Rocket Panel;[11] this is shown by Table XV based on some Harvard photographic results.[10]

The trend is obvious and in the expected direction; the divergence is in the ratio of 2.6 to one at $H = 95.5$ km. If the Rocket Panel densities above 90 km are underestimated, as suggested by some recent data,[20] the discrepancy becomes worse. While here the order of magnitude is still maintained, at greater altitudes the presence of dustballs makes the application of the deceleration method completely illusory.

TABLE XV

Comparison of Atmospheric Densities, Derived From Decelerations, With Those From the Rocket Panel

H, km	$\log \rho$ from decelerations	$\log \rho$ from rocket panel	$\Delta \log \rho$
60.0	− 6.40	− 6.46	+ 0.06
67.3	− 6.81	− 6.85	+ 0.04
71.5	− 7.14	− 7.11	− 0.03
79.8	− 7.44	− 7.66	+ 0.22
85.4	− 7.93	− 8.06	+ 0.13
89.3	− 8.12	− 8.34	+ 0.22
95.5	− 8.36	− 8.77	+ 0.41

For micrometeors, in which air and vapor molecules have free access to, or can freely leave the surface, and which are isothermal, $\zeta = 1$, $K = 1$, and $\zeta K = 1$.

CHAPTER 5

Ablation

5.1 Types and Heats of Ablation

Ablation of the meteoroid may proceed by way of *1*, vaporization, *2*, fusion and spraying of the liquid, *3*, sputtering, *4*, fragmentation, and *5*, by a combination of these processes.

Vaporization is the most important one, being the final stage of all types of ablation; fusion into one single drop, or spraying into many drops or fragments is followed by their vaporization.

Nevertheless, spraying of the fluid into small drops is a distinct type of ablation; the rate of ablation of the meteoroid in this case is determined by the heat of fusion. Melting into *one* drop, however, involves only a transformation of shape, and is not a stage of ablation.

Sputtering as a process of ablation in the earth's atmosphere is unimportant except for micrometeors of high velocity.

Fragmentation is common among meteorites. Fragmentation under aerodynamic pressure of liquid drops shed by the meteoroid is another case. Fragmentation of *dustballs* under very low drag, of the order of 10^4 d/cm^2, results in a great number of small, probably solid, fragments each of which is a micrometeor in its own right.

For solid meteoroids the type of ablation depends upon the size, velocity, conductivity of heat, viscosity of the melt, air cap, and rotation. With increasing size the succession is as follows: *1*, no ablation for slow, and ablation by sputtering for fast micrometeors, without melting; *2*, micrometeors melting into one drop and subsequently partially evaporating, with energy dissipation mainly through radiation; *3*, meteors melting into one drop and almost completely evaporating; *4*, same as the preceding, but the drop breaking up into minor fragments which evaporate; *5*, ablation through fusion and spraying of the liquid; *6*, partial spraying with vaporization from the liquid surface; *7*, vaporization "on the spot," without noticeable spraying. The limiting radii for each type depend upon the physical constants of the material and upon rotation, and are greater for iron than for stone. For ordinary stone meteors types *3* and *7* probably prevail, whereas for iron meteors *3* and *5* are to be chiefly considered. Mixed types of ablation must be quite common.

The limiting conditions for each type of ablation are summarized in Section 6.11.

For an initial temperature of 200° K the heats of ablation can be estimated as follows.

Nickel-iron, heating of solid to 1780° K (temperature of fusion), $h_1 = 1.09 \times 10^{10}$ (erg/g); heat of fusion $h_2 = 0.27 \times 10^{10}$; heating of liquid to 2500° K (median temperature of evaporation), $h_3 = 0.48 \times 10^{10}$; heat of vaporization $h_4 = 6.42 \times 10^{10}$; heat of ablation by fusion and spraying (at temperature of fusion),

$$h_f = h_1 + h_2 = 1.36 \times 10^{10} \text{ (erg/g)}; \tag{5–1}$$

heat of preheating from 200° to 2500° K (temperature of vaporization for visual meteors),

$$h_p = h_f + h_3 = 1.84 \times 10^{10} \text{ (erg/g)}; \tag{5–2}$$

heat of ablation by vaporization on the spot,

$$h_v = h_p + h_4 = 8.26 \times 10^{10} \text{ (erg/g)}. \tag{5–3}$$

Stone, heating of solid to 1800° K (temperature of fusion), $h_1 = 1.43 \times 10^{10}$ (erg/g); heat of fusion, $h_2 = 0.27 \times 10^{10}$; heating of liquid to 2100° K (temperature of vaporization for visual meteors), $h_3 = 0.33 \times 10^{10}$; heat of vaporization, $h_4 = 6.05 \times 10^{10}$; heat of ablation by fusion and spraying (at 2100° K),

$$h_f = h_p = h_1 + h_2 + h_3 = 2.03 \times 10^{10}\,(\text{erg/g}); \quad (5\text{–}4)$$

heat of ablation by vaporization on the spot,

$$h_v = h_f + h_4 = 8.08 \times 10^{10}\ (\text{erg/g}). \quad (5\text{–}5)$$

For stone, the high viscosity will not allow spraying at fusion; with heating the melt becomes more fluid, but even at 2100° K, the estimated temperature of vaporization, the viscosity of olivine is about 12 poise; spraying under these conditions is possible only at fast rotation,[6] at the same time vaporization will be intense, so that pure spraying of the melt will never take place in stone.

Dustballs crumble into fragments of the order of $\leqq 0.03$ cm radius[30] (Leonid, stone), or 0.005–0.010 cm (Draconids 1946), which undergo ablation by melting into one drop and then evaporating.

5.2 Normal and Abnormal Environment

The atmospheric environment will be called normal if it is reached by the unbroken individual meteoroid on a continuous flight from interplanetary space. An abnormal environment is that encountered by a meteoroid injected into its surroundings without the prehistory of continuous flight; fragments and drops, dustgrains of dustballs and artificial meteors will find themselves in abnormal environment.

As an approximation, we will assume here the model of a *homotropic* atmosphere, i.e., one built up according to an exponential law with constant adjusted scale height. Deceleration will also be neglected.

Ablation (depth ablated, Δr) will be assumed proportional to the kinetic energy of the air mass per cm² cross section traversed by the meteor. In normal environment and in Problem 1 ($\gamma \cong$ const) this is

$$\Delta r \sim w^2 \sec Z \cdot \exp\,(-H/a) \sim \rho w^2 a \sec Z. \quad (5\text{–}6)$$

The total ablation of the meteoroid is measured by the average thickness $r_0\delta/B$; statistically, for average conditions, this is proportional to r_0 (initial radius). Hence the density of air at the point of disappearance of an average meteor is

$$\rho_0 \sim r_0 \cos Z/(w^2 a). \tag{5-7}$$

5.3 The Typical Light Curve

The luminosity of a meteor is proportional to the mass of vapors evolved per unit time, or to the rate of ablation. Its variation along the trail defines the *light curve* of the meteor. For homotypic ablation, i.e., when there is only one phase, either vaporization, or fusion with spraying "on the spot," the form of the light curve can be derived as follows. For a given meteor we define a *parameter of ablation,*

$$X = \Delta r/r_0 = \rho/\rho_0 = \exp\left[-(H - H_2)/a\right], \tag{5-8}$$

according to Eqs. 5–6, 5–7 and the homotropic model of the atmosphere; H_2 is the altitude of the end point. For constant velocity the intensity of vaporization is proportional to ρr^2, where

$$r = r_0 - \Delta r = r_0(1 - X). \tag{5-9}$$

Hence the intensity of ablation, and thus also the relative luminosity will be

$$j = X(1 - X)^2 r_0^3 \cos Z \times \text{const}, \tag{5-10}$$

where the constant factor depends on w, δ, and atomic composition of the meteoroid. Equation 5–10 defines the typical light curve and is equivalent to that derived by Hoppe.[2]

Table XVI describes the variation of luminosity and other parameters along the typical trail. The maximum of luminosity is at $X = \frac{1}{3}$, $j = \frac{4}{27}$, when $\frac{2}{3}$ of the radius and $\frac{8}{27}$ of the meteor mass survive. The last line of the table gives the relative luminosity expressed in stellar magnitudes. The motion of the meteor is from bottom to top in the table. In the 5th column of the table, y is the length of the trajectory, in units of $a \sec Z$; with the length as abscissa, the luminosity is represented by a skew curve rising gradually and declining

TABLE XVI

Model of Homotypic Ablation in Normal Environment and Homotropic Atmosphere

Relative radius $(1 - X)$, area $(1 - X)^2$, mass $(1 - X)^3$, and luminosity (j) along trajectory. y = linear length counted backwards from end point in units of $a \sec Z$.

$X = \rho/\rho_0$	$(1-X)$	$(1-X)^2$	$(1-X)^3$	$y=(H-H_2)/a$	j/j_{max}	Δ mag (+)
1	0.000	0	0	0	0	∞
0.9	0.100	0.010	0.001	0.10	0.061	3.04
0.8	0.200	0.040	0.008	0.22	0.216	1.66
0.7	0.300	0.090	0.027	0.36	0.425	0.93
0.5	0.500	0.250	0.125	0.69	0.848	0.18
0.333	0.667	0.444	0.296	1.10	1.000	0.00
0.2	0.800	0.640	0.512	1.61	0.864	0.16
0.15	0.850	0.722	0.614	1.90	0.729	0.34
0.10	0.900	0.810	0.729	2.30	0.547	0.66
0.05	0.950	0.902	0.857	3.00	0.305	1.29
0.02	0.980	0.960	0.941	3.91	0.130	2.22
0.01	0.990	0.980	0.970	4.60	0.066	2.95
0.005	0.995	0.990	0.985	5.30	0.033	3.69

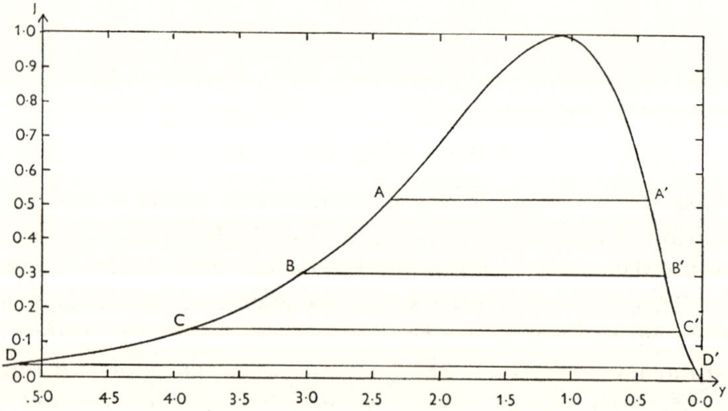

Fig. 8. Typical light-curve of a meteor in homotypic ablation. Abscissa, $y = (H - H_2)/a$, is the length of the path in units of $a \sec Z$ (about 10 km on the average). Ordinate, j, is the relative luminosity. The chords represent the fraction of the trajectory observable with the naked eye: AA′, 4th magnitude, BB′, 3rd magnitude; CC′, 1.5 magnitude; DD′, zero magnitude meteor.

steeply in a manner that is typical for meteor light curves (Fig. 8).

5.4 Observable Range and Length of Trail

The extent to which a meteor can be seen along the trajectory depends upon its brightness; for the same relative light curve, bright meteors are noticed earlier and fade later than the faint ones. From the experience of meteor observations, threshold differences of luminosity, Δ mag, and other visibility data are estimated as given in Table XVII.

TABLE XVII

Conditions of Visibility of Meteor Trail for Typical Light Curve

Apparent magnitude (average over observed trail), $\mathfrak{M}$	4.0	3.7	3.1	2.4	1.5	− 0.1
Faintest magnitude noticeable (at appearance or disappearance)	4.5	4.5	4.0	3.5	3.0	2.5
Magnitude at maximum	3.8	3.4	2.7	2.0	0.9	− 1.1
Threshold Δ mag (*min-max*)	+ 0.7	+ 1.1	+ 1.3	+ 1.5	+ 2.1	+ 3.6
Threshold j/j_{max}	0.52	0.36	0.30	0.25	0.14	0.036
X_e	0.66	0.73	0.76	0.78	0.84	0.92
$(H_a - H_d)/a = \Delta$	1.9	2.4	2.8	3.0	3.6	5.1

The last line of Table XVII gives the theoretical range in observable height (for normal environment and vaporization on the spot), in units of the scale height, calculated with the assumed Δ mag from the 5th and 7th columns of Table XVI. Here H_a = height of observed appearance, H_d = height of observed disappearance. $\Delta = y_a - y_d$, where y_a and y_d are the two end-point values of y from the 5th column of Table XVI, each corresponding to Δ mag of Table XVII. In the 6th line of Table XVII, $X_e = \rho_d/\rho_0$ is the parameter of ablation for the point of *observed* disappearance.

In Table XVIII observed ranges in height from the Arizona Expedition[42] are compared with those calculated for the typical light curve from the factors of the last line of Table XVII. The scale height is assumed according to Table I. In the table, "average magnitude" is the observed average apparent magnitude of the group, corrected for observational error-dispersion.

TABLE XVIII

Visual Meteors (Sporadic 80 %, Shower 20 %), Arizona
Observed and Calculated Range in Height

(the material consists mostly of dustballs)

Aver. magnitude, $\mathfrak{M}$	4.0	3.7	3.1	2.4	1.5	−0.1	All magnitudes
Group I: radiants* < 57° from apex; $\bar{w} \cong 68$ km/sec							
$\bar{H}$, km	98.2	102.2	101.6	99.6	106.2	98.0	—
n	94	109	242	177	135	95	—
$H_a - H_d$, obs., km	9.9	12.5	17.0	15.1	19.3	27.2	—
$H_a - H_d$, calc., km	13.3	17.3	20.2	21.0	27.3	35.2	—
Ratio obs/calc.	0.74	0.72	0.84	0.72	0.71	0.77	0.750
Group II: radiants* 57°–90° from apex; $\bar{w} \cong 42$ km/sec							
$\bar{H}$, km	92.2	96.0	91.3	92.4	91.2	85.2	—
n	135	178	331	244	128	70	—
$H_a - H_d$, obs., km	8.9	11.8	12.2	16.4	17.2	23.7	—
$H_a - H_d$, calc., km	12.3	16.3	18.2	19.5	23.4	32.1	—
Ratio obs/calc.	0.72	0.72	0.67	0.84	0.74	0.74	0.738
Group III: radiants* > 90° from apex; $\bar{w} \cong 21$ km/sec							
$\bar{H}$, km	87.6	89.5	89.0	88.1	83.6	84.4	—
n	101	140	300	206	88	56	—
$H_a - H_d$, obs., km	7.6	9.6	9.6	12.4	14.2	25.5	—
$H_a - H_d$, calc., km	11.9	15.4	17.9	19.2	22.3	31.6	—
Ratio obs/calc.	0.64	0.62	0.54	0.65	0.64	0.81	0.650
All groups							
Average obs/calc.	0.70	0.69	0.68	0.74	0.70	0.77	0.713

* "Radiant" is the point defining the direction of motion of an individual meteor on the celestial sphere.

Table XVIII is arranged into velocity groups; $\bar{H} =$ arithmetical mean height above sea level of center of trail (equal to the observed harmonic mean *plus* 2.5 km as a correction for cosmic dispersion); $n =$ observed number of meteors; $H_a - H_d =$ difference between height of appearance and disappearance, observed and calculated.

The 5th line in each group gives the ratio of observed to calculated range. There is no definite variation of this ratio with magnitude, but a marked decrease with decreasing velocity, indicating different physical conditions for the slow meteors moving in direct orbits (Group III) as compared with the rest. The average observed range is 71 per cent of that calculated on the vaporization-on-the-spot model. Although the result depends upon the assumed model of ablation, the smaller range in height may be related to the dustball nature of these meteors. The dustgrains are released in abnormal environment, at a greater atmospheric density than they would have reached as independent bodies, and their visible trajectories are therefore shorter. However, as will be seen in Section 5.8, the type of ablation of dustball grains actually requires a range in height equal to 0.75 of that in homotypic ablation, as compared to an observed ratio of 0.713. The agreement is almost perfect and indicates that the grains of dustballs are practically in normal environment. This can be understood if it is assumed that the dustball dissolves into grains before appreciable heating sets in, and that all the grains are already detached at the beginning of the visible trail.

Let L be the visible length of the trail, and let us inquire into the statistical ratio of length to range in height. We have

$$H_a - H_d = L \cos Z, \text{ or } L = (H_a - H_d) \sec Z. \qquad (5\text{–}11)$$

In normal environment the range is fixed in units of the scale height (Table XVII), and L is variable. The frequency of incidence from Z to $Z + dZ$ is proportional to $\cos^2 Z \sin Z \, dZ$, where one $\cos Z$ factor corresponds to luminosity (Eq. 5–10; a decrease in luminosity in a given ratio approximately corresponds to an equal decrease in observable meteor numbers). Hence the weighted average length of the trail, for random incidence and normal environment, becomes

$$\bar{L} = 1.5(H_a - H_d). \tag{5–12}$$

In the extreme case of abnormal environment of great density $L = \text{const}$ and $H_a - H_d$ is variable; there is no dependence of luminosity upon incidence and the frequency of incidence is proportional to $\cos Z \sin Z \, dZ$; the weighted average range becomes

$$\bar{H}_a - \bar{H}_d = \tfrac{2}{3}L, \tag{5–13}$$

which is identical with Eq. 5–12. It is thus safe to assume that, whatever the environment, in random incidence the average ratio of length to range will be 1.5.

5.5 Velocity-Mass Relation

A relation can be established between the ablating mass and the decelerating velocity of the meteoroid, which depends only on the coefficients of transfer and is valid irrespective of shape, aspect, and environment.[1]

The kinetic energy of air swept by the meteoroid in unit time is $\frac{1}{2}\sigma\rho w^3$, where σ is the *instantaneous* cross section. Of this a fraction γ is spent on ablating the meteoroid, and the decrement of mass becomes

$$\frac{dM}{dt} = -\frac{\gamma\sigma\rho w^3}{2h}. \tag{5–14}$$

The aerodynamic drag is $\zeta K\sigma\rho w^2$, and the deceleration

$$\frac{dw}{dt} = -\frac{\zeta K\sigma\rho w^2}{M}. \tag{5–15}$$

Time, cross section, and atmospheric density can be eliminated by dividing Eq. 5–14 by Eq. 5–15; the result is

$$\frac{dM}{M} = \frac{\gamma w \, dw}{2\zeta K h}; \tag{5–16}$$

the equation is exact for any instant; γ is the *net* coefficient of transfer, i.e., the difference between kinetic transfer and loss through radiation; and h is the net heat of ablation, or the sum of the heat of ablation proper (vaporization, fusion) *plus* that simultaneously spent on heating the meteoroid.

When there is no ablation, $h = \infty$ and Eq. 5–16 becomes useless. During heating without ablation, with q as the heat content per unit mass of the meteoroid, instead of Eq. 5–14 we have

$$\frac{dq}{dt} = +\tfrac{1}{2}\gamma\sigma\rho w^3. \tag{5–17}$$

This together with (5–15) yields

$$dq = -\frac{\gamma w\,dw}{2\zeta K}. \tag{5–18}$$

Equations 5–16 and 5–18 define the relation between mass, heat content, and velocity in quite a general manner. The coefficients of the equations may vary during the flight, but the equations can be integrated either numerically, or over a range for which the coefficients are reasonably constant. Equation 5–16 yields

$$\ln(M_2/M_1) = \Gamma(w_2^2 - w_1^2)/h, \tag{5–19}$$

where $\Gamma = \gamma/(4\zeta K)$. Similarly, from (5–18)

$$q_2 - q_1 = \Gamma(w_1^2 - w_2^2). \tag{5–20}$$

Table XIX contains results of calculations according to Eqs. 5–19 and 5–20 for five schematic cases. For distinct values of the surviving mass ratio, M/M_1, it gives the calculated velocity, w. The last column contains M_0/M_1, the residual mass ratio. M_1 is the original mass, M_0 the residual mass of the kernel which survives when the velocity is reduced to zero. Actually, ablation ceases at a finite velocity, and the residual mass will be slightly greater than calculated.

From Sections C and D of Table XIX it appears that for meteorites of about 50 cm original diameter the residual

TABLE XIX

Deceleration and Ablation in Typical Cases

M/M_1 (w, km/sec)						
1.000	1.000	0.500	0.100	0.008	Γ	M_0/M_1
A. Bright stony meteor (photographic); compact; vaporization on the spot; ζK from Table XIV at $d = 1$; $\gamma = 0.6$, radiation losses neglected; $h = 8.08 \times 10^{10}$						
12.00	12.00	10.82	7.14	—	0.2000	0.0283
30.00	30.00	28.92	26.22	21.48	0.0872	6×10^{-5}
60.00	60.00	58.96	56.32	51.99	0.0435	4×10^{-9}
B. Bright iron meteor (photographic); compact; fusion and spraying; $\zeta K = 1$; $\gamma = 0.6$, radiation losses neglected; $h = 1.36 \times 10^{10}$						
12.00	12.00	11.75	11.10	10.01	0.1500	1×10^{-7}
30.00	30.00	29.90	29.65	29.26	0.1500	10^{-43}
60.00	60.00	59.95	59.83	59.64	0.1500	10^{-172}
C. Stony meteorite, $r_0 \cong 25$ cm; vaporization on the spot; $\zeta K = 0.5$; $\gamma = 0.1$; $h = 8.08 \times 10^{10}$						
12.00	12.00	5.67	—	—	0.0500	0.410
15.00	15.00	10.63	—	—	0.0500	0.248
20.00	20.00	16.97	5.29	—	0.0500	0.0895
25.00	25.00	22.65	15.91	—	0.0500	0.0231
30.00	30.00	28.07	22.98	10.98	0.0500	0.0044
D. Iron meteorite, $r_0 \cong 25$ cm; mixed spraying and vaporization; $\zeta K = 0.5$; $\gamma = 0.1$; $h = 5 \times 10^{10}$						
12.00	12.00	8.64	—	—	0.0500	0.238
15.00	15.00	12.48	—	—	0.0500	0.105
20.00	20.00	18.18	13.04	—	0.0500	0.0183
25.00	25.00	23.57	19.87	11.93	0.0500	1.92×10^{-3}
30.00	30.00	28.82	25.88	20.42	0.0500	1.23×10^{-4}
E. Stony grain of dustball; complete fusion followed by vaporization; $\zeta K = 1$; $\gamma = 0.8 - 4.4 \times 10^5/w$, allowance being made for loss by radiation. Two-stage process: preheating without ablation, $q_2 - q_1 = 2.03 \times 10^{10}$; subsequent vaporization, $h = 6.05 \times 10^{10}$						
12.00	11.19	9.37	—	—	0.1083	0.107
30.00	29.79	29.37	28.48	26.62	0.1633	9×10^{-11}
60.00	59.91	59.72	59.26	58.55	0.1817	10^{-47}

fraction of mass decreases rapidly with velocity; also, for stone the fraction is greater than for iron. For more than one-thousandth of the mass to survive, the velocity must be less than 27 km/sec for iron, and 34 km/sec for stone; for more than 10 % to survive, $w < 15$ km/sec for iron and 19 km/sec for stone. For larger meteorites the surviving fraction will be greater, first because of the smaller value of the coefficient of heat transfer ($\gamma \leqq 0.01$), and second because they can penetrate the atmosphere without complete loss of momentum.

Section E of Table XIX gives the deceleration during ablation of a typical dustgrain. In this case losses by radiation become important because the event takes place in atmospheric strata of low density. From observations[8, 9, 43, 44] it appears that, regardless of velocity, dustballs of all sizes dissolve into grains at an aerodynamic drag per unit cross section of about 10^4 d/cm^2; the average drag along their trajectory may be about double that value, or $\rho w^2 \cong 2 \times 10^4$. With $\kappa = 0.8$ the impact heat transferred to the meteoroid becomes $\frac{1}{4} \cdot \frac{1}{2}\kappa \rho w^3 = 0.1 \rho w^3$ per cm^2 of *surface*. At a vaporization temperature of 2100° K the loss from black-body radiation is 1.10×10^9 (erg cm^{-2} sec^{-1}), and the net gain is $\frac{1}{4} \cdot \frac{1}{2}\gamma \rho w^3 = -1.1 \times 10^9 + 0.1 \rho w^3$. Substituting the value of ρw^2 as estimated above, we obtain

$$\gamma = 0.8 - 4.4 \times 10^5 w^{-1}, \qquad (5\text{–}21)$$

which includes radiation losses and is valid for the average environment of the dustgrains.

For very small micrometeors radiation losses may balance the heat supply at temperatures well below those of efficient evaporation. Such objects will have $\gamma = 0$ and will be completely decelerated without losing much mass. However, for higher velocities sputtering comes into play.

The rate of sputtering is given in Eqs. 4–66 and 4–67; the ratio of sputtered to incident mass can be represented as

$$\Theta = k[(w^2/c^2) - 1], \qquad (5\text{–}22)$$

with $k = 1.77$, $c = 42 \times 10^5$ for iron, and $k = 1.56$, $c = 51 \times 10^5$ for stone. The equation of mass, instead of Eq. 5–14, now becomes

$$\frac{dM}{dt} = -\Theta\sigma\rho w. \tag{5–23}$$

With Eqs. 5–15, 5–22 and $\zeta K = 1$, after integration from $w = w$ to $w = c$, the residual mass ratio for sputtering becomes

$$\ln (M_0/M_1) = \tfrac{1}{2}k[1 - (w/c)^2 + \ln (w/c)], \tag{5–24}$$

where w is the initial velocity.

Table XX gives the calculated residual mass ratios for sputtering.

TABLE XX

Residual Mass Ratio for Sputtering

w, km/sec	$\leqq 42$	51	$\leqq 51$	60	72	100
Iron, M_0/M_1	= 1	0.778	—	0.546	0.289	0.035
Stone, M_0/M_1	= 1	1	1	0.843	0.601	0.184

5.6 General Equations of Ablation and Motion

For most meteors, except those of very oblique incidence, the curvature of the earth (i.e., of atmospheric equidensity strata), the variation of gravity across the atmosphere, and the curvature of the meteor path itself can be neglected. In such a case, with an accuracy better than 0.1 per cent,

$$A = p \sec Z/g, \tag{5–25}$$

where A = air mass (g/cm^2) at altitude H per unit of meteor cross section, reckoned from the top of the atmosphere along the meteor path; Z = zenithal angle of incidence; p = atmospheric pressure at H; g = acceleration of gravity at an altitude $H + a$.

For an element of path $dl = -dH \cdot \sec Z$, $dA = \rho dl$, or

$$dA = -\rho dH \sec Z. \tag{5–26}$$

The general equation of ablation can be written as

$$dM = -\tfrac{1}{2}\gamma\sigma w^2 \cdot dA/h, \tag{5–27}$$

where M = mass, γ = net coefficient of heat transfer (deducting radiation), h = gross heat of ablation (net heat of vaporization and/or fusion *plus* that spent in heating the meteoroid), σ = instantaneous cross section. With these definitions formula 5–27 is exact and can be used as a basis in numerical integrations. In most problems average values of σ in random orientation, and the corresponding values of γ would suffice. The dependence of γ, σ, and h upon M and A has to be defined for that purpose.

With $dw/dt = w\,dw/dl$ and $dl = dA/\rho$, Eq. 5–15 transforms into

$$dw = -\zeta K\sigma w\,dA/M; \tag{5–28}$$

here ζK must also be defined as depending upon M and A.

The differential equations 5–27 and 5–28 are the basis of all calculations referring to the theory of ablation and deceleration. The equations are simple, but complications can be great in allowing for the various factors involved, which are generally variable along the path of the meteor.

From Eqs. 4–1, 4–3, and 4–6, $dM = (3\sigma/BC)\delta\,dr$. When this is equated to Eq. 5–27, σ cancels out and

$$dr = -BC\gamma w^2 dA/(6h\delta). \tag{5–29}$$

This is again a generally valid formula, with BC, γ, and h generally variable along the path. Similarly, with $\sigma/M = BC/(r\delta)$, which follows from Eqs. 4–1, 4–3, and 4–6, Eq. 5–28 is quite generally transformed into

$$dw = -(\zeta K)(BC)w\,dA/(r\delta). \tag{5–30}$$

Equations 5–29 and 5–30, replacing Eqs. 5–27 and 5–28,can also be used conveniently in the calculations.

5.7 The Standard Model

The standard model of meteor ablation, used for approximate calculation of the physical parameters along the meteor path, is based on Eq. 5–29 in which constant or average values of the parameters are assumed; as a first approximation, the velocity is also assumed constant. The standard model is successfully applied only to those small meteors which have no air cap ($d \leqq 2$), and is used also to define the limits of its own validity.

In Eq. 5–29, with γ, w, δ, and h assumed constant, an average value of BC remains to be defined. This factor is not only variable from meteor to meteor, but, generally, it may considerably vary along the meteor path. A solid rotating meteoroid experiences nonuniform ablation, depending upon the orientation of its rotational axis with respect to the direction of motion. Therefore both B and C will vary in a nonrandom fashion along the trajectory, and the value of the product, averaged over the meteor trail, will be called the *coefficient of ablation*,

$$J = \tfrac{1}{3}(BC)_{\text{Av}}. \qquad (5\text{–}31)$$

The simplest case is that of an evaporating spherical drop for which $B = 3$, $C = \frac{1}{4}$, and $J = \frac{1}{4}$.

For a given initial distribution of the shape parameters an average value of J for solid meteoroids can be obtained by considering the variation of B and C during ablation. We assume conventionally ablation "on the spot" and rectangular prismatic shape. The faces of the prism will wear down according to how they are exposed to the impact of the air stream. This depends upon the position of the axis of rotation of the meteoroid and the angle which this axis makes with the direction of flight. Another conventional simplification consists in assuming uniform ablation of each face, so that a prism remains a prism.

Thus, a cube moving parallel to its axis of rotation and at right angles to one face will wear down only from the front side; its cross section will remain constant and its thickness decrease; in the beginning it is a cube, $B = 3.72$, $C = \frac{1}{6}$, $\frac{1}{3}BC = 0.207$; at the end it becomes a square leaf of zero thickness, $B = \infty$, $C = \frac{1}{2}$, $\frac{1}{3}BC = \infty$; averaging by integration, with the element of thickness as weight, leads to a mean value of $J = 0.620$.

More detailed calculations[8] for a distribution of shape as in Table III, Case II, yielded $J = 0.44$ for a random orientation of the axes of rotation of different meteors with respect to the direction of motion. For the average solid meteoroid of not too unusual shape we assume

$$J = 0.5 \pm 0.1.$$

Treating J as a constant, integration of Eq. 5–29 yields the equation of ablation of the standard model,

$$r_0 - r = J\gamma w^2(A - A_0)/(2h\delta). \tag{5-32}$$

Here r_0 is the initial radius, A_0 the air mass at the beginning of ablation, or end of preheating.

For the phase of preheating without ablation, the absorbed heat is $dq = \frac{1}{2}\gamma_0(\sigma/M)w^2\,dA$; with $\sigma/M = BC/(r\delta) = 3J_0/(r_0\delta)$, after integration the total heat of preheating, or that absorbed per unit mass until the start of ablation becomes

$$q = 3J_0\gamma_0 w^2 A_0/(2r_0\delta). \tag{5-33}$$

Here J_0 and γ_0 refer to the stage of preheating.

Substituting A_0 from Eq. 5–33 and r from Eq. 5–9 into Eq. 5–32, we arrive at the basic equation of the standard model, valid for arbitrary environment,

$$A = 2[q/(3J_0\gamma_0) + hX/(J\gamma)]r_0\delta/w^2. \tag{5-34}$$

This is a relation between two variables, X, the parameter of ablation (which through Eq. 5–9 defines the radius), and A, the total air mass traversed. The equation allows for a two-phase process, ablation starting after a certain amount of preheating. When $q = 0$, it transforms into the case of homotypic ablation without preheating.

5.8 Atmospheric Density-Radius Relation in Standard Model

Let us first consider ablation in normal environment and a homotropic atmosphere, when $\rho = \rho_0 \exp(-H/a)$;

integration of Eq. 5–26 then yields

$$\rho = A \cos Z/a. \tag{5–35}$$

An effective value of the scale height should be chosen, namely that corresponding to Eqs. 5–25 and 5–35, $a = p/(\bar{g}\rho)$, where $\bar{g}$ is the gravity at $H + a$. With $p = NkT$, $\rho = N\mu$, this becomes $a = kT/(\bar{g}\mu)$, which differs from the local scale height Eq. 2–2 by the factor $g/\bar{g} \cong 0.998$, or practically unity.

Replacing A in Eq. 5–35 by Eq. 5–34, we obtain an expression for the atmospheric density as a function of the parameter of ablation, X, in normal environment:

$$\rho = \phi r_0 \cos Z/w^2, \tag{5–36}$$

$$\phi = 2[q/(3J_0\gamma_0) + hX/(J\gamma)]\delta/a. \tag{5–37}$$

The formula is almost independent of the assumed structure of the atmosphere except for the local scale height; for approximate calculations an average value of $a = 6.5 \times 10^5$ cm has been here assumed; this corresponds to $H = 90$ km which is very close to the average height of visual meteor trails.

Equation 5–36 can also be used to calculate the initial equivalent radius, r_0, when the altitude, H, and thus ρ at a certain phase of ablation is known from observation.

In the following, eight typical cases of the application of Eq. 5–36 and 5–37 are considered. The results are of a statistical nature only, individual deviations (e.g., on account of shape) being considerable. Losses by radiation are neglected.

Case 1. Isothermal iron meteoroid in normal environment, melting into one drop and then evaporating.

Case 2. Isothermal stony meteoroid in normal environment, melting into one drop and then evaporating.

Case 2a. Nonisothermal stony meteoroid, melting into one drop through the action of fast rotation.

Cases 1 and 2 apply also in the case of nonisothermal meteoroids in which the liquid is swept to the rear by drag and collected into one drop.

Case 3. Isothermal iron meteoroid in normal environment, spraying the liquid at fusion.

Case 4. Nonisothermal iron meteoroid in normal environment, spraying the liquid at fusion "on the spot."

Case 5. Nonisothermal stony meteoroid in normal environment, spraying the liquid at fusion "on the spot."

Case 5a. Spraying at fusion after isothermal heating of the solid stony meteoroid in rapid rotation.

Case 6. Nonisothermal iron meteoroid in normal environment, vaporizing "on the spot."

Case 7. Nonisothermal stony meteoroid in normal environment, vaporizing "on the spot."

TABLE XXI

Adopted Constants and Density Factor, ϕ, for the Typical Cases of Standard Model Ablation in Homotropic Atmosphere (normal environment, $a = 6.5 \times 10^5$ cm)

Case	δ	q	h	$J_0\gamma_0$	$J\gamma$	ϕ
1	7.8	1.84×10^{10}	6.42×10^{10}	0.200	0.150	$7.36 \times 10^5(1+13.9X)$
2, 2a	3.4	2.03×10^{10}	6.05×10^{10}	0.278	0.200	$2.55 \times 10^5(1+12.4X)$
3	7.8	1.09×10^{10}	2.70×10^9	0.500	0.260	$1.74 \times 10^5(1+1.47X)$
4	7.8	0	1.36×10^{10}	—	0.260	$12.55 \times 10^5 X$
5	3.4	0	2.03×10^{10}	—	0.390	$5.45 \times 10^5 X$
5a	3.4	1.43×10^{10}	6.00×10^9	0.500	0.390	$1.00 \times 10^5(1+1.62X)$
6	7.8	0	8.26×10^{10}	—	0.300	$66.08 \times 10^5 X$
7	3.4	0	8.08×10^{10}	—	0.300	$28.20 \times 10^5 X$

Table XXI contains the constants and the expression for ϕ for these cases.

In Table XXII numerical values of ϕ are given for five characteristic points on the trail of a typical ablating meteoroid: F, the flash point, or beginning of ablation, and end of preheating; B, when 0.1 of the radius and 0.273 of the mass has been ablated, and when the trail has become quite bright (effective beginning); C, the effective middle of the trail; D, when 0.5 of the radius and 0.875 of the mass has been ablated, the brightness being still high but rapidly declining (effective end point); E, the true end point, when ablation is complete.

It must be remembered that these results are obtained by neglecting deceleration and radiation losses and are, therefore, not valid for the smaller micrometeors of low velocity in which ablation is insignificant. On the other hand, Table XXII should more or less apply to dustball grains of velocities exceeding 16 km/sec.

TABLE XXII

Atmospheric Density Factors in Normal Environment, Standard Model, $a = 6.5 \times 10^5$ cm

Point	F	B	C	D	E	
r/r_0	1.000	0.900	0.750	0.500	0.000	
X	0.000	0.100	0.250	0.500	1.000	
j/j_{max}	0.000	0.547	0.950	0.848	0.000	
Case	$\phi = \rho w^2 \sec Z/r_0$ in 10^5 units					l $(B - D)$
1	7.36	17.6	33.0	58.5	109.6	1.20
2(2a)	2.55	5.72	10.5	18.4	34.2	1.20
3	1.74	2.00	2.38	3.02	4.30	0.41
4	0	1.26	3.14	6.28	12.6	1.61
5	0	0.54	1.36	2.72	5.45	1.61
5a	1.00	1.16	1.41	1.81	2.62	0.44
6	0	6.61	16.5	33.0	66.1	1.61
7	0	2.82	7.05	14.1	28.2	1.61

The last column of Table XXII lists the relative length of path between the characteristic points B and D, computed from $l = \ln (\rho_D/\rho_B)$. We note that in Cases 1 and 2 (with preheating) the length of path is 0.75 that of Cases 4, 5, 6, 7 (ablation "on the spot"). Reverting to Table XVIII we find that the observed lengths of trajectories of visual meteors are in almost the same ratio (0.65 to 0.75) to those calculated with ablation "on the spot." The suggestion is very near that the dustball grains, responsible for the phenomenon of visual meteors, actually ablate according to the preheating model, or Case 2, a suggestion which appears plausible in view of the small size of the grains rendering them isothermal.

For definitely abnormal environment, when the atmospheric density is high and the length of path small, the density will vary little along the trail and can be assumed constant. Such is the case with droplets or fragments shed by the meteoroid; these fragments are released at greater atmospheric densities than those which they would have reached in separate flight from space. Instead of Eq. 5–35 we can write in such a case

$$A = \rho L, \tag{5–38}$$

where L is the length of the path of ablation. Substituting this into Eq. 5–34, we obtain for the standard model of abnormal environment an equation, to be used instead of Eq. 5–36 when $L \ll a \sec Z$:

$$\rho = 2[q/(3J_0\gamma_0) + hX/(J\gamma)]r_0\delta/Lw^2. \tag{5–39}$$

CHAPTER 6

Details of Energy Transfer and Types of Ablation

6.1 Air Cap

As a working model, the formation of an air cap in front of a spherical meteoroid will be considered here. Within a close order of magnitude, the conclusions may be also valid for nonspherical shape. (Fig. 9).

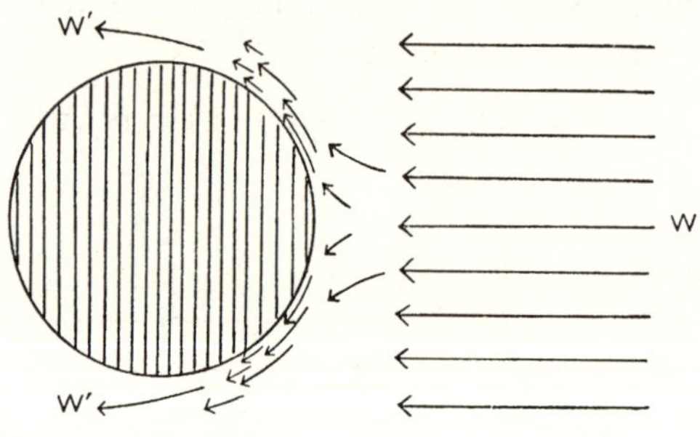

Fig. 9. Air cap formation in front of spherical meteoroid. Incident air stream of velocity w. Air cap stream leaves with velocity w'.

The density in the air cap is greater than in the free atmosphere (Eq. 4–38), and it can be shown that the linear thickness of the cap is small in comparison with the radius. In such a case the flow of mass out of the air cap can be set equal to $2\pi r \alpha w'$, where α (g/cm^2) is the mass of the cap per cm^2 of meteoroid surface, and w' the velocity of flow in the cap. The inflowing mass is $\pi r^2 \rho w$. Equating these two expressions, with $w' \cong \frac{1}{2}w$, we obtain as an order-of-magnitude estimate $\alpha \cong r\rho$.

In a spherical model of meteorite ablation (Öpik, to be published)

$$a = \tfrac{3}{4} r\rho \qquad (6\text{–}1)$$

has been derived by a more refined method for the effective surface mass of the cap. This will be adopted here. The effective kinetic depth of the air cap is then

$$d = \tfrac{3}{4} r\rho/(\lambda\rho) = \tfrac{3}{4} r/\lambda; \qquad (6\text{–}2)$$

here λ is the half-energy range in the free atmosphere at about one-half the velocity of the meteor. We note that λ, and thus d may greatly differ, according to velocity. The kinetic depth with respect to the incident molecules (at meteor velocities) will be smaller than that for the molecules of the air cap itself and the meteor vapors (thermal velocities). An incipient air cap, transmitting a considerable fraction of the incident kinetic energy, may be impenetrable to its own molecules.

TABLE XXIII

Half-Energy Air Mass, $\lambda\rho$, for Nitrogen in Nitrogen

Velocity, km/sec	$\lambda\rho$, 10^{-8} g/cm^2	$\Sigma(\lambda\rho)$, 10^{-8} g/cm^2
0.58	1.3	1.3
0.82	1.7	3.0
1.17	1.9	4.9
1.65	2.5	7.4
2.34	3.1	10.5
3.30	3.9	14.4
4.68	4.8	19.2
6.61	5.8	25.0
9.37	7.0	32.0
13.22	8.4	40.4
18.7	8.9	49.3
26.4	14.2	63.5
37.5	14.2	77.7
52.9	13.8	91.5
74.9	13.6	105.1
105.9	13.9	119.0
149.8	25.1	144.1

From a theory of interaction between atoms, cross sections and energy dissipation in elastic collisions have been derived that appear to be in satisfactory agreement with experiments (Section 7.4). For collisions of nitrogen in nitrogen (molecular or atomic) the figures as given in Table XXIII have been calculated.

The second column gives $\lambda\rho$, the air mass which reduces the velocity $\sqrt{2}$ times or to the nearest smaller tabular value; this is the half-energy air mass. Thus, at $w = 26.4$ km/sec an air mass $\lambda\rho = 14.2 \times 10^{-8}$ g/cm^2 (11 times the gaskinetic value) would reduce the velocity to $w = 18.7$ km/sec. The third column of the table contains the cumulative air mass, required to reduce a given velocity to the room-temperature molecular velocity; at $w = 26.4$ this is 63.5×10^{-8} g/cm^2, or about 50 times the gaskinetic "free path."

From Table XXIII we infer that $\lambda\rho \cong 14 \times 10^{-8}$ for $w > 25$ km/sec, and $\lambda\rho \cong 8.5 \times 10^{-8}$ for $12 < w < 20$ km/sec. Hence, from Eq. 6–2,

$$d = \psi r \rho, \tag{6–3}$$

where $\psi = 5.4 \times 10^6$ for $w > 25$ km/sec, and $\psi = 8.8 \times 10^6$ for $12 < w < 20$ km/sec.

With Eqs. 5–9 and 5–36, Eq. 6–3 for normal environment is transformed into

$$d = \phi\psi(1 - X)r_0^2 \cos Z/w^2. \tag{6–4}$$

For the middle point C, with $(\cos Z)_{Av} = \frac{2}{3}$, $X = 0.25$, and ϕ_c to be taken from Table XXII, Eq. 6–4 becomes

$$\bar{d} = \tfrac{1}{2}\phi_c \psi r_0^2/w^2. \tag{6–5}$$

The transition from Problem 1 to Problem 2 can be set effectively at $\bar{d} = 2$. For this case Eq. 6–5 yields the limiting radii as given in Table XXIV. The "cases" are those of Table XXII.

From heat conductivity and viscosity considerations,[6] it appears that air caps are of significance only in Cases 4, 6, and 7. Also, the figures of Table XXIV are well above the radii of compact meteors of the visual range (0.1–0.5 cm). For these and, of course, for dustball grains and micrometeors the air cap is practically absent.

TABLE XXIV

Limiting Radii for Incipient Air Cap ($\check{d} = 2$), Normal Environment

w, km/sec	12	20	25	40	60
			r_0, cm		
Case 1	0.44	0.74	1.2	1.9	2.9
Case 2	0.80	1.3	2.1	3.3	5.0
Case 3	1.6	2.8	4.4	7.1	10.6
Case 4	1.4	2.4	3.8	6.1	9.2
Case 5	2.2	3.7	5.8	9.3	14.0
Case 6	0.63	1.0	1.7	2.7	4.0
Case 7	0.96	1.6	2.6	4.1	6.2

6.2 Deformation of Drops and Limit of Spraying by Drag

A liquid drop of equivalent radius r will flatten out under aerodynamic drag into a figure resembling an oblate spheroid, with the minor semi-axis b in the direction of motion (Fig. 10). The results of an approximate analysis of the equilibrium conditions are summarized below. Let c be the

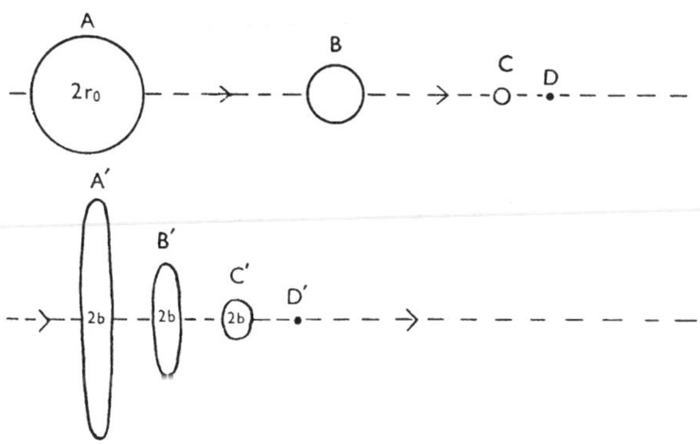

Fig. 10. Path of ablation of spherical drop (ABCD), and foreshortened path of flattened (or broken up) drop of equal mass (A′B′C′D′).

major semi-axis, s the surface tension, and p_s the drag per unit cross section. There are two equilibrium solutions, either

$$b = 3s/p_s \text{ (for flattened shape)}, \qquad (6\text{–}6)$$

or

$$r = b = c \leqq 4s/p_s = r_s \text{ (for roundish shape)}. \qquad (6\text{–}7)$$

Also, from the definition of equivalent radius,

$$\tfrac{4}{3}\pi c^2 b = \tfrac{4}{3}\pi r^3, \text{ or } b = r(b/c)^{\frac{2}{3}}. \qquad (6\text{–}8)$$

Appreciable flattening begins when the radius exceeds r_s as given by Eq. 6–7; an almost discontinuous transition to Eq. 6–6 then takes place. The result of flattening is an increased cross section and surface area. At considerable flattening the drop may break into smaller ones on account of uneven distribution of pressure, tangential drag and vibration. From observations of bubbles in water (Ö.) it appears that breakup almost certainly happens at $b/c = \frac{1}{4}$; from Eq. 6–8, $b = 0.4r$ and from Eq. 6–6 the condition of probable breakup becomes

$$r > r_b = 7.5s/p_s = 1.9r_s. \qquad (6\text{–}9)$$

The fragments have an average ellipticity of $b/c = \frac{1}{2}$ (observed, Ö.), $b = 0.625r$ (Eq. 6–8), and the average equivalent radius of the fragments, according to Eq. 6–6, becomes

$$r_f = 4.8s/p_s = 1.2r_s. \qquad (6\text{–}10)$$

There are thus three distinct phases of deformation.

1. *Stable phase.* When $r < r_s$ as defined by Eq. 6–7, surface tension prevails and the drop retains its spherical shape with little distortion.

2. *Deformation phase.* When $1.9r_s > r > r_s$, the drop flattens out into an oblate pseudo-spheroid with $b = \frac{3}{4}r_s$, according to Eqs. 6–6 and 6–7, and with an average load per unit cross section inversely proportional to drag,

$$M/\sigma = r_s\delta = 4s\delta/p_s \text{ (g/cm}^2\text{)}. \qquad (6\text{–}11)$$

3. *Breakup phase.* When $r > 1.9r_s$, the spheroid will hardly survive and will break up into smaller pieces with an average equivalent radius of $r = r_f = 1.2r_s$ (Eq. 6–10); the load, given by Eq. 6–11, remains exactly the same as in the deformation phase. As the course of ablation depends mainly on the load, it matters little whether a drop breaks up or not.

In the case of flattened drops, the principle of random orientation is no longer valid. Nevertheless, in the absence of an air cap the drag formula will not differ much from Eq. 4–32 with $K = 1$, or from the second expression of Eq. 4–24, and the coefficient of heat transfer $\gamma \simeq \kappa$ will not differ much from its spherical or random equivalent. With an air cap, such as is the case with ultra-speed pellets in a dense atmosphere, the drag coefficient K for flattened drops will be nearer to 1 than to 0.5, so that the drag formula will be practically the same as without an air cap.

The simplifications of the standard model are not well applicable to flattened drops. Shape and contour coefficients are to be calculated directly from the cross section, πc^2, and the surface area, S, as defined by

$$S = \pi c^2 \left[2 + \frac{(1 - e^2)}{e} \ln \frac{(1 + e)}{(1 - e)} \right], \tag{6–12}$$

$$e^2 = 1 - b^2/c^2 = 1 - [3r_s/(4r)]^3, \tag{6–13}$$

$$c^2 = 4r^3/(3r_s). \tag{6–14}$$

For normal environment and standard model, Eq. 5–36 gives $p_s = \rho w^2 = \phi r_0 \cos Z$. Hence, according to Eqs. 6–7 and 5–9,

$$r/r_s = \phi(1 - X)r_0^2 \cos Z/(4s). \tag{6–15}$$

For middle trail $X = 0.25$, $\phi = \phi_c$ (Table XXII) and, with $(\cos Z)_{Av} = 2/3$ as an average,

$$r/r_s = \phi_c r_0^2/(8s). \tag{6–16}$$

This equation may be used to define the limits of strict

validity of Cases 1 and 2 of the standard-model approximation. These cases postulate spherical drops without deformation, or $r/r_s \leqq 1$. Equation 6–16 gives then, for Case 1, $s = 1200$, $\phi_c = 3.30 \times 10^6$,

$$r_0 < 0.054 \text{ cm (iron drop)}, \tag{6–17}$$

and for Case 2, $s = 360$, $\phi_c = 1.05 \times 10^6$,

$$r_0 < 0.052 \text{ cm (stone drop)}. \tag{6–18}$$

Special estimates indicate that deformation and breakup of liquid drops will not be efficiently hindered by viscosity, not even in the case of stone.

It may seem that drag distortion may offer an alternative to the dustball model, to explain the greatly increased ablation cross section of visual meteors. This question is considered in Section 6.4.

As to spraying of the melt from the surface of the meteoroid, this depends on viscosity as well as on surface tension. Here we consider only the second factor. When spraying is due to drag, and the fluid is moving freely along the surface, drops which are smaller than about the limit set by Eq. 6–7 cannot be detached; also the droplets must be small as compared with the meteoroid, whence $r/r_s \gg 1$ and, from Eq. 6–16,

$$r_0^2 \gg 8s/\phi_c. \tag{6–19}$$

This defines the lower limit of radius, for spraying by drag to be possible. Table XXV contains the result (for normal environment).

TABLE XXV

Minimum Radius for Spraying by Drag (Capillarity Criterion)

	Case 3	Case 4	Case 5
	(Iron, isothermal)	(Iron, nonisothermal)	(Stone, nonisothermal)
$r_0 \gg$, cm	0.20	0.18	0.15

6.3 Conductivity and Temperature Differences During Preheating

In this and the following sections some results of a consideration of greatly simplified models of heat transfer are briefly summarized.

In a solid nonrotating meteoroid there is a flow of heat from front to rear. During the phase of preheating, including fusion, radiation and evaporation can be neglected, and the conducted heat is spent on heating the meteoroid; the temperature gradient is greatest at the front and vanishes at the rear. The average thickness is $4r/B$ (Eq. 4–10), the temperature gradient at the front, equal to double the average, is $\frac{1}{2}\Delta T \cdot B/r$, and the front flow of heat per unit cross section is $\frac{1}{2}\gamma \rho w^3$. In such a case, with k_t denoting thermal conductivity, the difference of temperature between the two sides is close to

$$\Delta T = \Delta T_0 = \gamma \rho w^3 r/(Bk_t) \text{ (nonrotating).} \qquad (6\text{–}20)$$

In the presence of equalizing rotation (Section 4.2), the flow of heat is from the surface inwards. The effective difference of temperature over average thickness, r/B (Eq. 4–5), and through an inner surface equal to about three times the cross section, is then about one-twelfth of that given by Eq. 6–20, or

$$\Delta T = \Delta T_r = \gamma \rho w^3 r/(12Bk_t) \text{ (rotating).} \qquad (6\text{–}21)$$

The average temperature of the meteoroid is close to

$$\bar{T} = T - \tfrac{2}{3}\Delta T. \qquad (6\text{–}22)$$

The above equations can be used to define the limits of isothermacy during preheating. For order-of-magnitude estimates, let practical isothermacy correspond to $\Delta T < 0.3T$. At fusion, $T \simeq 1800°$ K, thus $\Delta T < 540°$. With this and $X = 0$ as for the end of preheating, also $r = r_0$, $\phi = \phi_f$ (Table XXII, Point F), $\cos Z = \frac{2}{3}$, $B = 5$ and ρw^2 from Eq. 5–36, from Eqs. 6–21 and 6–20 the limits of validity of the isothermal Cases 1, 2, and 3 are calculated as given in Table

XXVI. For reasons put forward in Section 4.2, the "rotating" set of values in the table is probably prevailing.

TABLE XXVI

Upper Limits of Equivalent Radius for the Validity of the Assumption of Isothermal Melting (Normal Environment, Standard Model)

	Case 1. Iron			Case 2. Stone			Case 3. Iron		
	$\gamma=0.76, k_t=4\times10^6$			$\gamma=0.89, k_t=2\times10^5$			$\gamma=1.00, k_t=4\times10^6$		
w, km/sec	12	25	60	12	25	60	12	25	60
r_0 (rotating)	0.54	0.37	0.24	0.187	0.130	0.084	0.97	0.67	0.43
r_0 (nonrotating)	0.16	0.11	0.069	0.054	0.038	0.024	0.28	0.19	0.12

6.4 Deformation of Molten Meteoroids and Foreshortening of the Trail

Thus, according to the second line of Table XXVI, stony meteoroids smaller than 0.19–0.084 cm will melt completely before evaporation begins. However, the drops will flatten out and even may break up; the increased area will lead to quicker ablation (Fig. 10). The speed of ablation is inversely proportional to the load; this is given by Eq. 6–11, as compared with the spherical load of $\frac{4}{3}r_0\delta$ of the nondeformed drop. With $p_s = \rho w^2$ per unit cross section taken from Eq. 5–36, conventionally as for normal environment of the nondeformed sphere, and $\cos Z = \frac{2}{3}$, the load ratio of deformed to nondeformed drop becomes

$$y = 4.5s/(\phi r_0^2). \tag{6–23}$$

This is also the foreshortening ratio of the ablation path. When $y = 1$, or in the absence of deformation, $\phi = \phi_c$ (Table XXII) should be taken; when $y = 0$, or when the ablation path is infinitesimal, $\phi = \phi_f$ as for Point F is the proper value. As a rough approximation, we assume for the effective value of the density factor in Eq. 6–23 the linear expression $\phi = \phi_f + (\phi_c - \phi_f)y$. With this, Eq. 6–23 transforms into

$$r_0^2 y[\phi_f + (\phi_c - \phi_f)y] = 4.5s. \qquad (6\text{–}24)$$

With $\phi_f = 2.55 \times 10^5$, $\phi_c = 10.5 \times 10^5$ as for Case 2 (Table XXII), and $s = 360$ d/cm as for stone, the foreshortening ratios of Table XXVII are calculated. The validity of the

TABLE XXVII

Foreshortening of Ablation Path of a Completely Molten Stony Meteoroid

r_0, cm	≤ 0.052	0.052	0.061	0.084	0.139	0.191	0.224
Foreshortening ratio, y	1	0.72	0.60	0.40	0.20	0.125	0.100

table is limited to meteors which remain in normal environment until the end of fusion. At $r_0 = 0.052$ cm there is a discontinuity, Eq. 6–24 being superseded by the condition 6–18 of minimum deformation. The upper limit of radius is set by the condition of complete melting, or that of Table XXVI. Thus, foreshortening will actually take place within a range of radii of $0.052 < r < r_0'$, r_0' being the limit given in Table XXVI ("rotating"). Thus, at $w = 12$ km/sec the foreshortening varies from 1 to 0.125, as corresponding to the range of r_0 from 0.052 to 0.19 cm; at $w = 25$ km/sec the ratio is from 1 to 0.21 ($r_0 = 0.052$–0.13 cm); and at $w = 60$ km/sec it is from 1 to 0.40 ($r_0 = 0.052$–0.084 cm). The foreshortened trail will be brighter than for a meteor of equal mass in normal environment, and the deceleration abnormally high. The length of the trail will decrease with increasing brightness over the critical range of radius, a circumstance which may serve to decide whether the increase in cross section through flattening or fissure of the molten meteoroid is taking place. No such effect has been observed statistically for the visual meteors [cf. Table XVIII and Ref. 42) which should show it between the fourth and second magnitudes if they were compact bodies; on the contrary, the length of the trail of visual meteors steadily increases with luminosity. Therefore, the alternative explanation to the dustball model, that of the increased area of ablation as an effect of deformation of a drop, is not acceptable. Yet a few exceptional cases of short trails and "bursts" are probably due to compact iron and stone meteoroids, which appear to be rare among visual meteors.

6.5 Equalization of Evaporation and Radiation in the Absence of Rotation

The transport of heat from front to rear has a direct bearing on the relative intensity of evaporation in different directions, thus on the jet correction. In a nonrotating solid meteoroid the transport can be brought about only by conductivity. In liquid meteoroids convective transport may also play a significant role.

We consider a state after preheating, when a more or less steady temperature distribution in the molten meteoroid has been reached. The heat of aerodynamic friction is then spent on vaporization and radiation. A constant gradient of temperature, equal to $\frac{1}{4}\Delta T \cdot B/r$, establishes itself between front and rear and, instead of Eq. 6–20, the temperature difference becomes

$$\Delta T = 2\varepsilon\gamma\rho w^3 r/(Bk_t), \qquad (6\text{–}25)$$

where ε is the fraction of total heat transported to the rear. From Eq. 6–25, with $r = r_0(1 - X)$, $X = 0.25$ as for Point C (Table XXII), $\phi = \phi_c$, $\cos Z = \frac{2}{3}$, $B = 3$ as for a molten

TABLE XXVIII

Limiting Radii for Equalization of Evaporation by Conductivity in the Absence of Rotation

(normal environment and standard model)

	Case 1, Iron			Case 2, Stone		
w, km/sec	12	25	60	12	25	60
Practical equalization ($\Delta T < 40°$),	$r_0 \leqq 0.020$	0.013	0.009	0.007	0.005	0.003
Absence of equalization ($\Delta T > 280°$),	$r_0 \geqq 0.082$	0.057	0.035	0.030	0.021	0.013

sphere, and ρw^2 given by Eq. 5–36, the limiting radii of Tables XXVIII and XXIX are calculated. The first table refers to evaporation, the second to radiation. Two cases are considered: that of "practical equalization," $\frac{1}{2} > \varepsilon > 4/9$; and of "absence of equalization," $\varepsilon < \frac{1}{6}$. The adopted differences of temperature correspond to evaporation varying as the 15th power, and radiation as the 4th power of temperature.

It appears that evaporation can be equalized by conductivity only in micrometeors, even in the case of iron.

TABLE XXIX

Limiting Radii for Equalization of Radiation by Conductivity in the Absence of Rotation
(normal environment and standard model)

	Case 1, Iron			Case 2, Stone		
w, km/sec	12	25	60	12	25	60
Practical equalization ($\Delta T < 150°$)	$r_0 \leqq 0.038$	0.027	0.016	0.013	0.010	0.006
Absence of equalization ($\Delta T > 700°$)	$r_0 \geqq 0.13$	0.089	0.055	0.048	0.033	0.021

The limits for radiation are higher than those for evaporation, but still well below those of Table XXVI; meteoroids in which equalization takes place will indeed undergo complete melting, previous to ablation, as has been assumed here.

6.6 Equalization by Convection in a Drop

In a liquid drop convection is caused by the difference in aerodynamic pressure and skin drag. A model of convective transport, which is supposed to yield an approximation of a close order of magnitude, led to the following expression, in usual notations:

$$\Delta T = Fw\eta/r, \tag{6–26}$$

with $F = 34.4\varepsilon\gamma/(c_p\delta)$. Here ε is the fraction of total heat transported to the rear, 34.4 is a dimensionless factor, and η the coefficient of viscosity. With $\gamma = 0.6$, and $c_p = 1.1 \times 10^7$, $\delta = 3.4$ for stone, $c_p = 6.7 \times 10^6$, $\delta = 7.8$ for iron, the coefficient F is found as follows:

	Stone		Iron	
	Equalization	Absence of equalization	Equalization	Absence of equalization
	$\varepsilon = 0.4$	$\varepsilon = 0.17$	$\varepsilon = 0.4$	$\varepsilon = 0.17$
F	2.2×10^{-7}	8.8×10^{-8}	1.6×10^{-7}	6.4×10^{-8}

Equation 6–26 is independent of environment and also applies to artificial meteors. The constant F, however, depends on environment and is proportional to $\gamma/(K\delta c_p)$, where K is the drag coefficient, assumed here equal to 1.

The efficiency of the drag-versus-viscosity equalization in liquid drops increases with increasing radius. The limits of equalization calculated from Eq. 6–26 are contained in Table XXX. The viscosity for stone, $\eta = 10$, would correspond to a temperature of evaporation of 2100° K.

From the table it appears that convective equalization is complete in iron droplets of practically all sizes, and that, whenever other means of equalization fail, the droplet will be kept practically isothermal by drag convection alone.

As to stone, the minimum radii of equalization in evaporation all exceed the limit of deformation, $r = 0.052$ cm (Eq. 6–18); molten stone drops will never evaporate equally from front and rear. On the other hand, absence of equalization takes place only at radii which are well below the dustball limit. It appears that stone drops from the dustball grain to the maximum permissible size all have partial equalization of evaporation; it is never complete, yet never completely absent. Therefore, for stony meteors of the dustball type the deceleration factor ζK will be between 1 and the values in the 2nd line of Table XIV ($d = 0$, no air cap); tentatively its run is estimated as follows (for $r = 0.03$ cm, interpolated between the second and third columns of Table XXX, assuming $\zeta K = 1$ for equalization, and taking the value of Table XIV for absence of equalization):

w, km/sec	12	25	60
ζK, dustball grains	1.0	1.3	2.6

TABLE XXX

Convective Equalization of Temperature in Liquid Drops (normal environment and standard model; $\gamma = 0.6$; $K = 1$; $\eta = 10$ for stone and 0.01 for iron)

	Stone		Iron
	Equalization	Absence of equalization	Equalization
	Evaporation		
	$\Delta T < 40°$	$\Delta T > 280°$	$\Delta T < 40°$
w, km/sec	Limit of r, cm		
12	> 0.066	$< 3.8 \times 10^{-3}$	$> 5 \times 10^{-5}$
25	> 0.14	$< 8.0 \times 10^{-3}$	$> 1.0 \times 10^{-4}$
60	> 0.33	< 0.019	$> 2.4 \times 10^{-4}$
	Radiation		
	$\Delta T < 150°$	$\Delta T > 700°$	$\Delta T < 150°$
w, km/sec	Limit of r, cm		
12	> 0.018	$< 1.5 \times 10^{-3}$	$> 1.3 \times 10^{-5}$
25	> 0.037	$< 3.2 \times 10^{-3}$	$> 3 \times 10^{-5}$
60	> 0.088	$< 7.6 \times 10^{-3}$	$> 6 \times 10^{-4}$

The trend with velocity is considerable. Of course, some additional equalization will be due to conductivity, but the figures of Table XXVIII indicate that, in the absence of rotational equalization (Section 6.7), this source of equalization is insignificant.

6.7 Surface Heat Transport by Rotation

Some heat is stored in the surface layers of the meteoroid during exposure to the air stream; when transported to the rear by rotation, the absorbed heat is given away to evaporation and radiation. This sort of heat transport is important only when direct transport by conductivity is inadequate. In such a case, the storage of heat by conduction into an infinite wall is a convenient model.

The solution is given by a combination of Gaussian functions, or their integrals. For close order-of-magnitude estimates a simpler procedure can be used.

A surface heat action results in an increment of temperature $\Delta T = y$ depending upon the depth x, such that $y(dy/dx) < 0$, and $y \to 0$ for $x \to \infty$. In such a case the integral of heat content per unit surface can be represented as

$$\int_0^\infty y\,dx = ky_0^2(-y_0'), \qquad (6\text{–}27)$$

with $y = y_0$ and $dy/dx = y_0'$ at $x = 0$. k depends upon the form of $y = f(x)$ as follows:

$y = f(x)$	Linear	Parabola	e^{-x}	$\int_x^\infty e^{-x^2}\,dx$
k	$\frac{1}{2}$	$\frac{2}{3}$	1	0.637

The last case of the table is an exact solution of the problem for constant surface temperature. It differs little from the parabola (which, however, requires $y = 0$ at finite x), and it appears to be legitimate to set $k = \frac{2}{3}$ for all cases of surface heating, irrespective of the actual boundary condition.

In the rotating meteoroid, one boundary condition is that of constant surface flux of energy. Consider an originally isothermal solid absorbing a flux Q (erg/cm^{2}sec) during an interval of time t. The total amount of absorbed heat, in notations of (6–27), will be

$$Qt = \tfrac{2}{3}\delta c_p y_0^2/(-y_0'). \qquad (6\text{–}28)$$

The conductive flux at the surface requires

$$-y_0' = Q/k_t. \qquad (6\text{–}29)$$

Equations 6–28 and 6–29 define the absorption of a constant flux of heat by an infinite, originally isothermal wall. In the present case, y_0 is the amplitude of the temperature variation at the surface during one rotation. When the surface is exposed to the air stream, its temperature rises at first steeply from a minimum value T_0 and then more slowly approaches the maximum of $T_0 + y_0$ (Fig. 11); the curve of y is concave

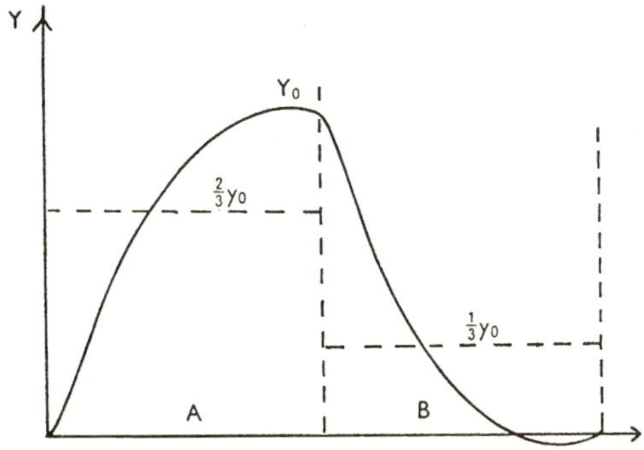

Fig. 11. Surface temperature variation with rotation. Abscissa, time; ordinate, temperature. A, half-period of rotation with air stream on (front); B, half-period with air stream off (rear).

to the t axis, and the average temperature of the front side is close to $T_0 + \frac{2}{3}y_0$. When the air stream is turned off, the temperature at first drops quickly from the maximum of $T_0 + y_0$ and then more slowly approaches the minimum of T_0, the y curve being convex to the t axis; the average temperature of the rear side is nearly $T_0 + \frac{1}{3}y_0$. The average difference of temperature between front and rear is thus $\frac{1}{3}y_0$. Setting $t = \frac{1}{2}P$, where P is the period of rotation, and substituting this and Eq. 6–29 into Eq. 6–28, with $y_0 = 3\Delta T$, we obtain

$$Q^2P = 12\delta c_p k_t(\Delta T)^2; \tag{6–30}$$

here ΔT is the average difference of surface temperature between front and rear.

If u is the "equatorial" velocity of rotation, $P \cong 6r/u$, then (6–30) yields

$$Q^2 r = 2u\delta c_p k_t(\Delta T)^2. \tag{6–31}$$

For stone, $\delta c_p k_t = 7.5 \times 10^{12}$; for iron, 2.1×10^{14} (cgs).

Equation 6–31 is valid for arbitrary environment. With $\gamma = 0.6$ its application to normal environment and evaporation requires:

for equalization, $\Delta T = 40°$,

$$Q = 0.4 \times \tfrac{1}{2}\gamma \rho w^3 \times \tfrac{1}{2} = 0.06\rho w^3;$$

for lack of equalization, $\Delta T = 280°$,

$$Q = 0.17 \times \tfrac{1}{2}\gamma \rho w^3 \times \tfrac{1}{2} = 0.0255\rho w^3.$$

The additional factor $\frac{1}{2}$ in Q allows for the average incidence over a hemisphere.

With $\cos Z = \frac{2}{3}$, ρw^2 from Eq. 5–36, $r = r_0(1 - X)$ and substituting the proper expressions for Q and ΔT into Eq. 6–31, the condition for rotational equalization of evaporation becomes

$$r_0^{\,3} = Nu/[(1 - X)\phi^2 w^2]. \tag{6–32}$$

The values of the coefficient N are as follows:

	Stone		Iron	
	Equalization	No equalization	Equalization	No equalization
$N =$	1.5×10^{19}	3.9×10^{21}	4.3×10^{20}	1.1×10^{23}

For $X = 0.25$, or Point C of Table XXII, the limiting radii are found from Eq. 6–32 as given in Table XXXI.

According to the table, rotational equalization turns out to be relatively inefficient. However, for molten stony dust-grains, $r < 0.03$ cm and $u \cong 10$ cm/sec (Section 4.2), practical equalization of evaporation is indicated and $\zeta K = 1$, contrary to nonrotating grains (Section 6.6).

TABLE XXXI

Limiting Radii for Rotational Equalization of Evaporation
(normal environment and standard model)

a. Stone

Case	2	2	7	7	7	7
	Isothermal		Nonisothermal		Nonisothermal	
u, cm/sec	10	10	100	100	1000	1000
	Equal.	No equal.	Equal.	No equal.	Equal.	No equal.
w, km/sec			r_0, cm			
12	< 0.050	> 0.32	< 0.14	> 0.90	< 0.31	> 2.0
25	< 0.031	> 0.20	< 0.087	> 0.55	< 0.18	> 1.2
60	< 0.017	> 0.11	< 0.049	> 0.31	< 0.11	> 0.66

b. Iron

Case	1	1	6	6	6	6
	Isothermal		Nonisothermal		Nonisothermal	
u, cm/sec	10	10	100	100	1000	1000
	Equal.	No equal.	Equal.	No equal.	Equal.	No equal.
w, km/sec			r_0, cm			
12	< 0.070	> 0.46	< 0.25	> 1.5	< 0.54	> 3.4
25	< 0.044	> 0.29	< 0.15	> 0.98	< 0.32	> 2.1
60	< 0.025	> 0.15	< 0.083	> 0.54	< 0.18	> 1.1

6.8 Radiation Losses and Temperature of Evaporation

The ratio of radiation to evaporation heat expenditure is

$$\Theta = Q/h\xi, \tag{6-33}$$

where Q is the radiation, ξ the evaporation, and h the effective heat of evaporation. The ratio determines the outcome of ablation: when Θ is small, evaporation prevails and ablation can proceed far toward its maximum possible limit; when it is large, the energy is chiefly dissipated through radiation, γ is small and ablation weak, as in micrometeors. Critical values of the ratio occur for small meteors which melt isothermally; hence $h \simeq 6 \times 10^{10}$ (erg/g) can be assumed as corresponding to evaporation of an already molten substance. With the expressions for ξ and vapor pressure as in Appendices 1 and 2, Q according to Eq. 4–61 with $\varepsilon = 1$ for stone, and according to Eq. 4–62 for molten iron, Table XXXII has been calculated. Q and $h\xi$ are given in erg/cm^2 sec. $Q + h\xi$ is the total amount of heat expended on radiation and evaporation per unit of area and time. $\Theta > 1$ and radiation dominates below 1800° K for stone, and 2100° K for iron. $\Theta < 0.1$ and radiation is insignificant above 2200° K for stone and 2500° K for iron.

TABLE XXXII

Radiation and Evaporation Losses and Their Ratio

T, °K	1600	1800	2000	2200	2400	2600	2800	3000
				Stone				
$\log Q$	8.570	8.775	8.958	9.124	9.275	9.414	9.543	9.662
$\log \xi$	−2.950	−2.039	−1.311	−0.718	−0.166	0.189	0.543	0.850
Θ	5.47	1.08	0.307	0.115	0.046	0.028	0.017	0.011
$\log (Q+h\xi)$	8.644	9.061	9.588	10.111	10.636	10.983	11.332	11.636
				Iron				
$\log Q$	7.878	8.116	8.329	8.522	8.697	8.859	9.009	9.147
$\log \xi$	−5.021	−3.777	−2.783	−1.961	−1.375	−0.871	−0.440	−0.066
Θ	124	12.2	2.03	0.475	0.184	0.084	0.044	0.025
$\log (Q+h\xi)$	7.881	8.149	8.502	9.013	9.505	9.970	10.384	10.751

For complete equalization of evaporation and radiation over the surface of the meteoroid, the gross flow, or the average amount of heat absorbed and expended per cm^2 of surface and second, is U/S; in the absence of equalization the heat turnover is confined to the front surface only, and the gross flow is $2U/S$. According to Eq. 4–25 this is

$$Q + h\xi = \tfrac{1}{8}\gamma\rho w^3 \tag{6–34}$$

for equalization, and

$$Q + h\xi = \tfrac{1}{4}\gamma\rho w^3 \tag{6–35}$$

in the absence of equalization.

In Table XXXIII equilibrium temperatures are given, calculated according to Eq. 6–34 and Table XXXII for complete equalization of evaporation. $\gamma = 0.8$ for stone and 0.6 for iron, and altitudes according to Table I have been assumed. In the absence of equalization the temperatures should be increased by 110–115° for small meteors melting isothermally and would refer to the front side of the meteor; and by about 70° for large meteors vaporizing "on the spot."

TABLE XXXIII

Equilibrium Temperature of Average Meteor Surface (isothermal case of "complete equalization")

w, km/sec		12	20	30	45	60	75
				Stone			
$\log[(Q + h\xi)/\rho] =$		17.237	17.903	18.431	18.959	19.334	19.625
H, km	$\log \rho$		T, °K				
80	− 7.676	1990	2244	2469	2772	3014	>3200
90	− 8.389	1698	1972	2174	2375	2578	2745
100	− 9.065	<1600	1693	1916	2117	2245	2371
110	− 9.684	<1600	<1600	1649	1881	2024	2135
120	−10.249	<1600	<1600	<1600	1632	1809	1920
130	−10.720	<1600	<1600	<1600	<1600	<1600	1725
				Iron			
$\log[(Q + h\xi)/\rho] =$		17.112	17.778	18.306	18.834	19.209	19.500
H, km	$\log \rho$		T, °K				
80	− 7.676	2372	2664	2934	>3200	>3200	>3200
90	− 8.389	2102	2353	2577	2833	3043	>3200
100	− 9.065	1724	2083	2293	2513	2684	2828
110	− 9.684	<1600	1759	2047	2256	2409	2534
120	−10.249	<1600	<1600	1731	2032	2179	2297
130	−10.720	<1600	<1600	<1600	1774	1993	2109

We will consider now an application of the tables to the theory of dustballs. Observations have shown[42, 8, 9, 43, 44] that, irrespective of brightness (i.e., total mass), velocity, and angle of incidence, meteors of the visual range become first visible at an altitude where $\rho w^2 = \text{const} \cong 10^4$ (d/cm²) when ρ is taken from the Rocket Panel.[11] The corresponding air mass is too small to cause evaporation of a compact meteoroid of the required size. The appearance at an altitude where the air mass is inadequate has been interpreted as due to a sudden increase of the cross section, caused by the breakup of a dustball into dustgrains when drag reaches a critical value. The amount of heat absorbed per cm^2 since entry into the atmosphere is proportional to $\rho w^2 \sec Z$, and the absence of the expected correlation,[42] $\rho \sim \cos Z$, rules out the possibility of an alternative process of ablation.

There is another possibility. The dustball may have broken up into grains before reaching the apparent flash point; however, sufficient light will be emitted only when noticeable evaporation starts, or when $Q + h\xi$ is high enough for that purpose. From Eq. 6–34 it appears that $\rho w^3 =$ const would satisfy this condition, being also independent of the angle of incidence. The figures of Table XXXIV,[8] containing the material of Table XVIII supplemented by some photographic observations, show how well the two rival hypotheses agree with the observations.

It appears that $\rho w^2 =$ const is a better hypothesis than $\rho w^3 =$ const; in the first case the extremes are in a ratio of 2.5 to 1, in the second the ratio is 6 to 1.

TABLE XXXIV

Flash Point Data

	Arizona visual meteors			Photogr. Draconids[45]	Photogr. Leonid[46]
	Fast	Medium	Slow		
w, km/sec	68	42	21	23	72
$\log (\rho w^2)$	4.02	4.26	4.01	3.87	4.13
$\log (\rho w^3)$	10.85	10.88	10.33	10.23	10.98

The second hypothesis can also be put to trial upon an absolute basis. Substituting into Eq. 6–34 (as for equalization, cf. Section 6.7) $\gamma = 0.8$ as for stone, $Q + h\xi = 0.1\rho w^3$, and with ρw^3 as from the last line of Table XXXIV, with the aid of Table XXXII, Table XXXV is obtained.

From the table it appears that Θ is always small, and that evaporation losses, far from being negligible, prevail.

TABLE XXXV

Temperature and Radiation to Evaporation Loss Ratio in Visual and Photographic Meteors (Stone Dustgrains)

w, km/sec	68	42	21	23	72
		Flash point			
$\log (Q + h\xi)$	9.85	9.88	9.33	9.23	9.98
T, °K	2100	2111	1902	1864	2150
Θ	0.19	0.18	0.56	0.72	0.15
		Middle trail			
$\log (Q + h\xi)$	10.46	10.49	9.94	9.84	10.58
T, °K	2333	2344	2135	2096	2379
Θ	0.06	0.06	0.16	0.19	0.05

6.9 Drag-Induced Flow of the Melt

The mobility of the liquid layer on the front side of the meteoroid determines the type of ablation. When the liquid is removed before its surface becomes heated to temperatures of intense evaporation, spraying will be the dominant process. On the other hand, surface tension sets lower limits to the radius, for spraying to be possible without the help of rotation; for normal environment the limits are given in Table XXV which refers to slow rotation, such that it may be

TABLE XXXVI

Limits of Radius for Drag-Induced Spraying to be Considered (normal environment)

	Stone	Iron			Iron		
	Case 5	Case 3			Case 4		
Velocity km/sec	All	12	25	60	12	25	60
Limits of r, cm	>0.15	0.20–0.54	0.20–0.37	0.20–0.24	>0.54	>0.37	>0.24

unable to cause flow, although it may be fast enough to induce isothermal melting. From a comparison with Table XXVI, "rotating" case, the overall limits of radius for spraying by drag are found as given in Table XXXVI. The limits are those for which spraying is possible if the viscosity is sufficiently low. It is by no means implied that spraying will actually take place within these limits.

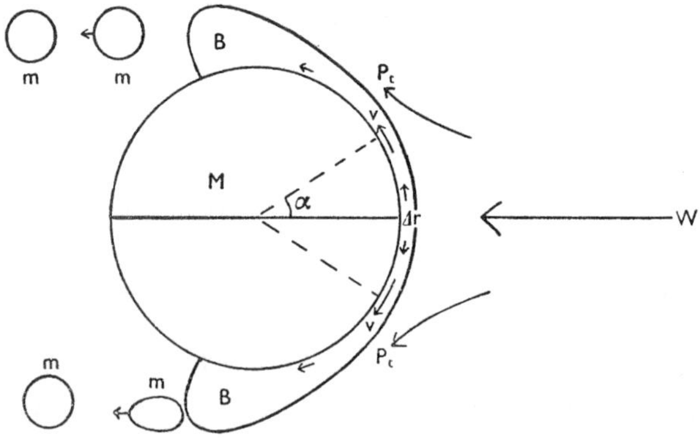

Fig. 12. Drag-propelled liquid layer (Δr) in front of spherical meteoroid (M). BB = marginal bulge; mm = detaching drops.

Drag pushes the liquid from front to rear. We consider a spherical model with a thin air cap;[6] as to order of magnitude, the results will be valid for arbitrary air caps (Fig. 12). In the notations of Section 4.3, the tangential component of drag is

$$p_t = (1 - q)\rho w^2 \cos \alpha \sin \alpha. \qquad (6\text{–}36)$$

The amount of melted material per sec over the front cap from $\alpha = 0$ to $\alpha = \alpha$ is $\frac{1}{2}\gamma \pi r^2 \rho w^3 \sin^2 \alpha / h_f$, where h_f is the effective heat of fusion. This equals the mass removed in viscous flow, $2\pi r v \delta \sin \alpha \cdot \Delta r$, where Δr is the thickness of the liquid layer at α, small as compared with r (if not small, the concept of flow fails), and v is the *average* velocity of flow. Hence, for complete removal of the melt by drag,

$$v \cdot \Delta r = \tfrac{1}{4}\gamma r \rho w^3 \sin \alpha / (h_f \delta). \qquad (6\text{–}37)$$

In viscous flow, the surface velocity is about double the average, and the gradient of velocity is $2v/\Delta r$. With η as the coefficient of viscosity, the shear per cm^2 becomes $2v\eta/\Delta r$. Equating this to Eq. 6–36 and combining it with Eq. 6–37, we get

$$\Delta r = \left[\frac{rw\eta\gamma}{2h_f\,\delta(1 - q)\cos\alpha}\right]^{\frac{1}{2}}, \tag{6–38}$$

$$v = \tfrac{1}{2}\rho w^2 \sin\alpha \left[\frac{rw\gamma(1 - q)\cos\alpha}{2\eta h_f\,\delta}\right]^{\frac{1}{2}}. \tag{6–39}$$

For $\alpha \to \pi/2$, $\Delta r \to \infty$, $v \to 0$. This means that the liquid accumulates into a bulge near the outer contour, whence it is blown away into spray.

For $d < 2$, $\gamma = \kappa = 1 - q^2$. For $\alpha =$ const, Table XXXVII shows the dependence of Δr and v upon the coefficient of accommodation.

TABLE XXXVII

Coefficient of Accommodation, Relative Velocity, and Layer Thickness in Viscous Drag Flow

q	1	0.8	0.632	0.447	0.2	0.0
$\gamma = \kappa$	0	0.36	0.60	0.80	0.96	1.00
$\Delta r \sim (1 + q)^{\frac{1}{2}}$	1.41	1.34	1.28	1.20	1.10	1.00
$v \sim (1 - q)(1 + q)^{\frac{1}{2}}$	0	0.27	0.47	0.66	0.88	1.00

Equations 6–38 and 6–39 are valid only when the accelerations can be neglected. The momentum per cm^2 is $v\Delta r\delta = \omega$ and its acceleration $d\omega/dt$ (from Eq. 6–37 with α as the variable) must be less than the driving force, p_t. With $d\alpha/dt = v/r$, the condition of validity becomes

$$v/\sin\alpha < v_0 = 4(1 - q)h_f/(\gamma w), \tag{6–40}$$

where v is given by Eq. 6–39. When this is not fulfilled, the velocity of flow can be assumed equal to

$$v' = v_0 \sin\alpha, \quad (v' < v), \tag{6–41}$$

v_0 being given by Eq. 6–40. Substituting v' instead of v into Eq. 6–37, the layer thickness in flow by free acceleration becomes

$$(\Delta r)' = \frac{\gamma^2 r \rho w^4}{16(1-q)h_f{}^2\delta}. \tag{6–42}$$

When $\Delta r > (\Delta r)'$, Δr or Eq. 6–38 is to be used; when not, Eq. 6–42 is the best approximation.

With laminar flow, the transport of heat is by conductivity, and the difference of temperature between the top of the liquid layer and the underlying solid surface becomes

$$\Delta T = (\tfrac{1}{2}\gamma \rho w^3 \cos \alpha \cdot \Delta r)/k_t. \tag{6–43}$$

The temperature of the surface of the liquid is

$$T_s = T_f + \Delta T, \tag{6–44}$$

where T_f = temperature of fusion. For efficient spraying T_s should be well below the temperature of vaporization. Taking $\alpha = 45°$, $\gamma = \kappa = 0.8$ as for an average point on the meteoroid, the preceding formulae are transformed into

$$\Delta r = [rw\eta/(h_f\delta)]^{\frac{1}{2}}, \tag{6–45}$$

$$\Delta r' = 0.072 r \rho w^4/(h_f{}^2\,\delta), \tag{6–46}$$

$$\Delta T = 0.28 \rho w^3 \cdot \Delta r/k_t, \tag{6–47}$$

which apply to meteoroids with a small air cap ($d < 2$), irrespective of environment.

In normal environment, with Eq. 5–36 and $\cos Z = \frac{2}{3}$, the equations are transformed into

$$\Delta r = [(1 - X) r_0 w\eta/(h_f\delta)]^{\frac{1}{2}}, \tag{6–48}$$

$$\Delta r' = 0.048\phi(1 - X) r_0{}^2 w^2/(h_f{}^2\,\delta), \tag{6–49}$$

$$\Delta T = 0.19\phi r_0 w \cdot \Delta r/k_t. \tag{6–50}$$

a. *Stone.* ΔT increases with the 1.5 power of r_0 and w. The lowest surface temperature, most favorable for spraying, obtains at the lowest possible values of $w = 1.2 \times 10^6$ and $r_0 = 0.15$ cm. With this, and $\cos Z = \frac{2}{3}$ in Eq. 5–36 (Case 5),

$\log \rho = -8.0$, which corresponds to an equilibrium temperature for vaporization and radiation of about 1840° (Table XXXIII); from Appendix 2, $\eta \cong 50$ for olivine, and probably much higher for the actual melt. Case 5 of Table XXII and middle trail, with $k_t = 2 \times 10^5$, $\delta = 3.4$, $h_f = 2.03 \times 10^{10}$, then yields $\Delta r = 2.33 \times 10^{-5}(r_0 w)^{\frac{1}{2}}$, and $\Delta T = 2.9 \times 10^{-6}(r_0 w)^{1.5}$. With r_0 and w as assumed, $\Delta T = 200°$, $T_s = 1800° + 200° = 2000°$; this exceeds the equilibrium temperature of vaporization. Evidently, even in this case the layer will not attain the required thickness and will begin boiling at the surface. It appears that, in normal environment, efficient spraying will not take place in stone under the influence of aerodynamic drag alone.

b. *Iron.* The conditions are very different, on account of small viscosity and high thermal conductivity. From a similar analysis as for stone, in normal environment and for middle trail, with $k_t = 4 \times 10^6$, $\delta = 7.8$, $h_f = 2.7 \times 10^9$ (Case 3) and 1.36×10^{10} (Case 4), $\eta = 0.021$, the following conclusions are reached. Within the limits of radius set by Table XXXVI for Case 3, or isothermal melting followed by spraying, $\Delta T = 1°$–$17°$ only and spraying will always take place.

In Case 4, the condition of viscous flow is $r_0 < 1.35 \times 10^7/w$. This covers practically what could be called "small meteors," $r_0 < 11$ cm ($w = 12$ km/sec) to $r_0 < 2$ cm ($w = 70$ km/sec). $\Delta T < 285°$ obtains, $T_s < 2085°$. This is low as compared with the temperatures of vaporization and, thus, within the limits of viscous flow spraying of iron melt is always efficient.

For free accelerated flow, Eq. 6–50 for iron under the assumed conditions becomes

$$\Delta T = 1.1 \times 10^{-19}(r_0 w)^3. \tag{6–51}$$

The temperature of efficient vaporization of iron is from 2300° to 2700° K (Table XXXIII). Thus $\Delta T < 400°$, $r_0 w < 1.6 \times 10^7$ can be assumed as the limit of efficient spraying by drag, and $\Delta T > 700°$, $r_0 w > 1.9 \times 10^7$ would correspond to prevalence of evaporation in free flow. The demarcation line is quite sharp, and closely links with the limit of validity of viscous flow. It follows that from $r_0 > 0.2$ cm (Table XXXVI) up to

$$r_0 < 1.6 \times 10^7/w, \tag{6–52}$$

iron meteors will be spraying even without the help of rotation.

6.10 Centrifugal Flow of the Melt

Rotation may condition spraying in smaller meteors when drag alone does not suffice. The effect depends on the orientation of the axis of rotation. We will consider here only the case of pole-on motion; this is the least favorable case for spraying, but the simplest to treat, as some of the equations will be in common with those of the preceding section; the angle of incidence, α, is in this case identical with the colatitude of rotation.

The meridional component of centrifugal acceleration directed equatorwards is $g_c = (u^2 \sin\alpha \cos\alpha)/r$, where u = equatorial velocity of rotation. The centrifugal drag per cm² is $p_c = g_c\delta \cdot \Delta r$. This is to be set equal to the viscous shear, $2v\eta/\Delta r$, whence $v/(\Delta r)^2 = \frac{1}{2}u^2\delta \cdot \sin\alpha \cos\alpha/(r\eta)$.

In combination with (6–37) this yields

$$(\Delta r)^3 = \gamma r^2 \eta \rho w^3/(2u^2\delta^2 h_f \cos\alpha), \tag{6–53}$$

$$v^3 = \gamma^2 u^2 r(\rho w^3)^2 \sin^3\alpha \cos\alpha/(32h_f^2\eta\delta), \tag{6–54}$$

valid under the condition of $v < u \sin\alpha$. Equation 6–47 for ΔT remains unaltered.

The liquid collects toward the equator of rotation, whence it is shed in drops; the average radius of a drop, r_d, can be estimated by equating the distending centrifugal force to capillarity cohesion, with s = surface tension; the result is

$$r_d = [3sr/(u^2\delta)]^{\frac{1}{2}}. \tag{6–55}$$

For true spraying the drops must be considerably smaller than the parent body; as an upper limit we set $r_d = \frac{1}{2}r$, whence the condition of validity of the concept of spraying becomes

$$r > 12s/(u^2\delta). \tag{6–56}$$

With $\gamma = 0.8$ and $\alpha = 45°$ as for an average point, the relations become

$$(\Delta r)^3 = 0.57r^2\eta\rho w^3/(u^2\delta^2h_f), \tag{6–57}$$

$$v^3 = u^2r(\rho w^3)^2/(200h_f^2\eta\delta), \tag{6–58}$$

when $v < u/\sqrt{2}$ and $\Delta r/r \ll 1$.

Further, for normal environment, $\cos Z = \frac{2}{3}$ and with Eq. 5–36 we obtain

$$(\Delta r)^3 = 0.38(1 - X)^2\phi r_0^3 w\eta/(u^2\delta^2h_f), \tag{6–59}$$

$$v^3 = u^2(1 - X)\phi^2 r_0^3 w^2/(450h_f^2\eta\delta), \tag{6–60}$$

with Eq. 6–50 for ΔT; Eqs. 6–59 and 6–60 are valid when

$$v < u/\sqrt{2} \text{ and } \Delta r/r < 0.5. \tag{6–61}$$

For stone, Case 5 of Table XXII and middle trail, with $\eta = 50$, $s = 360$ and other data as in Section 6.9,

$$\Delta r = 0.0183r_0(w/u^2)^{\frac{1}{3}}, \tag{6–62}$$

$$v = 7.6 \times 10^{-6}r_0(u^2w^2)^{\frac{1}{3}}, \tag{6–63}$$

$$\Delta T = 2.36 \times 10^{-3}r_0^2(w^4/u^2)^{\frac{1}{3}}, \tag{6–64}$$

for viscous flow which is practically always valid for the stony melt. The second condition Eq. 6–61 of validity (at $r = 0.75r_0$) is

$$w/u^2 < 8500, \tag{6–65}$$

and (6–56), with $r = 0.75r_0$, becomes

$$r_0 > 1690/u^2. \tag{6–66}$$

Assuming $\Delta T < 200°$ for the condition of efficient spraying of stone, the limiting radii are those of Table XXXVIII, calculated from Eq. 6–64.

TABLE XXXVIII

Upper Limits of Radius for Rotational Spraying of Stone
(normal environment; u = equatorial velocity of rotation)

u, cm/sec	0	10–100	300	1000
$r_0 > 1690/u^2$	∞	0.169	0.019	0.0017
w, km/sec		r_0 (cm),		
12	(0.054)	(0.187)	0.173 (0.187)	0.254
20	(0.042)	(0.145)	0.122 (0.145)	0.183
30	(0.034)	(0.119)	0.095 (0.119)	0.140
45	(0.028)	(0.097)	0.071 (0.097)	0.112
75	(0.022)	(0.075)	0.057 (0.075)	0.076

The bracketed cases are upper limits for complete fusion into one drop, as conditioned by rotation (Table XXVI), spraying being inhibited by capillarity ($u < 100$) or by size ($u = 300$); the figures without brackets are upper limits of r_0 for rotational spraying. Beyond the upper limits, vaporization on the spot (Case 7) sets in.

For iron the conditions are much simpler than for stone. The capillarity condition Eq. 6–56, with $r = 0.75r_0$ as for middle trail becomes

$$r_0 > 2470/u^2. \tag{6–67}$$

This is a sufficient and necessary condition for rotational spraying of iron meteoroids with $r_0 \leqq 0.20$ cm. For $r_0 > 0.20$ cm, spraying by drag supersedes that by rotation when $u < 10^4$ cm/sec, and Eq. 6–52 defines the upper limit of radius for spraying by iron, irrespective of rotation (unless rotation is unreasonably fast).

6.11 Charts of Types of Ablation

From a consideration of the phenomena determining ablation, and allowing for mutual exclusion or overlapping, Table XXXIX for stone, and Table XL for iron have been constructed. The "cases" or types of ablation are those of Section 5.1 and Table XXII. The basis for the classification of types is furnished by Tables XXVI, XXXVI, XXXVIII,

XXIV and Eqs. 6–52, 6–66, 6–67, and others. In the tables, u = equatorial velocity of rotation, r_0 = initial radius, w = velocity of the meteoroid; r_d = radius of droplets after fissure (r_f) or spraying (r_s).

The air cap limit in Tables XXXIX and XL corresponds to kinetic depth $d = 2$ (Table XXIV). The tables do not cover the case of thick air caps, and refer only to normal environment and average angle of incidence.

For $u \leqq 100$ cm/sec and purely aerodynamic disruption, the drop size after breakup of an unstable large drop, or for drag spraying, follows from Eq. 6–10, $p_s = \rho w^2$ and Eq. 5–36 as

$$r_d = r_f = 4.8s \sec Z/(\phi r_0). \qquad (6\text{–}68)$$

With $\sec Z = 1.5$ and $\phi = \phi_c$ as for middle trail (Table XXII), the drop size for $u \leqq 100$ and for stone is $r_d = r_f = 2590/(\phi r_0)$ (fissure); for iron, $u \leqq 100$, it equals $r_d = r_s = 8600/(\phi r_0)$ (spraying).

The drop size for rotational spraying, $u > 100$ cm/sec, as given in the second line of the table, is that defined by Eq. 6–55; however, if the resulting drop radius is greater than that given for $u \leqq 100$, the value for $u \leqq 100$ must be taken.

In Table XL, Case 1a is practically identical with Case 1, except that melting into one drop is achieved by drag flow, and not by conductivity.

The tables can be used to decide upon the type of ablation and for first-approximation calculations of the dimensions of a meteor or the physical conditions occurring along its trail, according to Table XXII and Eq. 5–36.

Although applying to normal environment, with sufficient precision the tables can also be used with respect to dustball grains. They help to set up a framework for more elaborate numerical integrations, according to the formulae of Section 5.6. Probable values of average rotation can be estimated as outlined in Section 4.2.

The tables refer to average conditions and can be applied with confidence only to statistical averages. Individual

TABLE XXXIX

Types of Ablation in Normal Environment, Compact Stone at Average Incidence and Shape

(The molten drops of Cases 2 and 2′ remain spherical for $r_0 < 0.052$, flatten out for $r_0 = 0.052$–0.072, and break up for $r_0 > 0.072$ cm) w = velocity of meteor, km/sec; u = rotational velocity, cm/sec; r_0 = initial equivalent radius, cm; r_d = droplet radius, cm.

u		= 0	10–100	200	300	1000
r_d		= $0.0025r_0^{-1}$	$0.0025r_0^{-1}$	$0.08r_0^{\frac{1}{2}}$	$0.05r_0^{\frac{1}{2}}$	$0.015r_0^{\frac{1}{2}}$
w						
12	Case 2, $r_0 <$	0.054	0.19	0.042	0.019	0.002
	Case 5a, $r_0 =$	—	—	0.042–0.15	0.019–0.17	0.002–0.19
	Case 5, $r_0 =$	—	—	—	—	0.19–0.25
	Case 2a, $r_0 =$	—	—	0.15–0.19	0.17–0.19	—
	Case 7, $r_0 >$	0.054	0.19	0.19	0.19	0.25
	Air cap, $r_0 >$	0.96	0.96	0.96	0.96	0.96
20	Case 2, $r_0 <$	0.042	0.15	0.042	0.019	0.002
	Case 5a, $r_0 =$	—	—	0.042–0.11	0.019–0.12	0.002–0.15
	Case 5, $r_0 =$	—	—	—	—	0.15–0.18
	Case 2a, $r_0 =$	—	—	0.11–0.15	0.12–0.15	—
	Case 7, $r_0 >$	0.042	0.15	0.15	0.15	0.18
	Air cap, $r_0 >$	1.6	1.6	1.6	1.6	1.6
30	Case 2, $r_0 <$	0.034	0.12	0.042	0.019	0.002
	Case 5a, $r_0 =$	—	—	0.042–0.083	0.019–0.095	0.002–0.12
	Case 5, $r_0 =$	—	—	—	—	0.12–0.14
	Case 2a, $r_0 =$	—	—	0.083–0.12	0.095–0.12	—
	Case 7, $r_0 >$	0.034	0.12	0.12	0.12	0.14
	Air cap, $r_0 >$	3.1	3.1	3.1	3.1	3.1
45	Case 2, $r_0 <$	0.028	0.097	0.042	0.019	0.002
	Case 5a, $r_0 =$	—	—	0.042–0.062	0.019–0.071	0.002–0.097
	Case 5, $r_0 =$	—	—	—	—	0.097–0.11
	Case 2a, $r_0 =$	—	—	0.062–0.097	0.071–0.097	—
	Case 7, $r_0 >$	0.028	0.097	0.097	0.097	0.11
	Air cap, $r_0 >$	4.6	4.6	4.6	4.6	4.6
60	Case 2, $r_0 <$	0.024	0.084	0.042	0.019	0.002
	Case 5a, $r_0 =$	—	—	0.042–0.055	0.019–0.063	0.002–0.084
	Case 5, $r_0 =$	—	—	—	—	0.084–0.090
	Case 2a, $r_0 =$	—	—	0.055–0.084	0.063–0.084	—
	Case 7, $r_0 >$	0.024	0.084	0.084	0.084	0.090
	Air cap, $r_0 >$	6.2	6.2	6.2	6.2	6.2
75	Case 2, $r_0 <$	0.022	0.075	0.042	0.019	0.002
	Case 5a, $r_0 =$	—	—	0.042–0.050	0.019–0.057	0.002–0.075
	Case 5, $r_0 =$	—	—	—	—	0.075–0.076
	Case 2a, $r_0 =$	—	—	0.050–0.075	0.057–0.075	—
	Case 7, $r_0 >$	0.022	0.075	0.075	0.075	0.076
	Air cap, $r_0 >$	7.8	7.8	7.8	7.8	7.8

TABLE XL

Types of Ablation in Normal Environment, Compact Iron at Average Incidence and Shape

(The molten drops of Cases 1 and 1′ remain spherical for $r_0 < 0.054$, flatten out for $r_0 = 0.054$–0.075, and break up for $r_0 > 0.075$ cm) w = velocity of meteor, km/sec; u = rotational velocity, cm/sec; r_0 = initial equivalent radius, cm; r_d = droplet radius, cm.

$u =$		0	10–100	200	300	1000
$r_d =$	Case 1	$0.0026r_0^{-1}$	$0.0026r_0^{-1}$	$0.09r_0^{\frac{1}{2}}$	$0.06r_0^{\frac{1}{2}}$	$0.019r_0^{\frac{1}{2}}$
	Case 3	$0.036r_0^{-1}$	$0.036r_0^{-1}$			
	Case 4	$0.027r_0^{-1}$	$0.027r_0^{-1}$			
w						
12	Case 1, $r_0 <$	0.16	0.20	0.050	0.022	0.002
	Case 1a, $r_0 =$	0.16–0.20	—	—	—	—
	Case 3, $r_0 =$	—	0.20–0.54	0.050–0.54	0.022–0.54	0.002–0.54
	Case 4, $r_0 =$	0.20–13	0.54–13	0.54–13	0.54–13	0.54–13
	Air cap, $r_0 >$	1.4	1.4	1.4	1.4	1.4
20	Case 1, $r_0 <$	0.12	0.20	0.050	0.022	0.002
	Case 1a, $r_0 =$	0.12–0.20	—	—	—	—
	Case 3, $r_0 =$	—	0.20–0.42	0.050–0.42	0.022–0.42	0.002–0.42
	Case 4, $r_0 =$	0.20–8	0.42–8	0.42–8	0.42–8	0.42–8
	Air cap, $r_0 >$	2.4	2.4	2.4	2.4	2.4
30	Case 1, $r_0 <$	0.098	0.20	0.050	0.022	0.002
	Case 1a, $r_0 =$	0.098–0.20	—	—	—	—
	Case 3, $r_0 =$	—	0.20–0.34	0.050–0.34	0.022–0.34	0.002–0.34
	Case 4, $r_0 =$	0.20–5.3	0.34–5.3	0.34–5.3	0.34–5.3	0.34–5.3
	Air cap, $r_0 >$	4.6	4.6	4.6	4.6	4.6
45	Case 1, $r_0 <$	0.080	0.20	0.050	0.022	0.002
	Case 1a, $r_0 =$	0.080–0.20	—	—	—	—
	Case 3, $r_0 =$	—	0.20–0.28	0.050–0.28	0.022–0.28	0.002–0.28
	Case 4, $r_0 =$	0.20–3.6	0.28–3.6	0.28–3.6	0.28–3.6	0.28–3.6
	Case 6, $r_0 >$	3.6	3.6	3.6	3.6	3.6
	Air cap, $r_0 >$	3.6	3.6	3.6	3.6	3.6
60	Case 1, $r_0 <$	0.069	0.20	0.050	0.022	0.002
	Case 1a, $r_0 =$	0.069–0.20	—	—	—	—
	Case 3, $r_0 =$	—	0.20–0.24	0.050–0.24	0.022–0.24	0.002–0.24
	Case 4, $r_0 =$	0.20–2.7	0.24–2.7	0.24–2.7	0.24–2.7	0.24–2.7
	Case 6, $r_0 >$	2.7	2.7	2.7	2.7	2.7
	Air cap, $r_0 >$	4.0	4.0	4.0	4.0	4.0
75	Case 1, $r_0 <$	0.062	0.20	0.050	0.022	0.002
	Case 1a, $r_0 =$	0.062–0.20	—	—	—	—
	Case 3, $r_0 =$	—	0.20–0.22	0.050–0.22	0.022–0.22	0.002–0.24
	Case 4, $r_0 =$	0.20–2.1	0.22–2.1	0.22–2.1	0.22–2.1	0.24–2.1
	Case 6, $r_0 >$	2.1	2.1	2.1	2.1	2.1
	Air cap, $r_0 >$	5.0	5.0	5.0	5.0	5.0

cases may show great divergences, especially at oblique angle of incidence, or for unusual shape and orientation.

As to Cases 2 and 1, the limits of validity as entered in Tables XXXIX and XL indicate that complete isothermal melting takes place; further ablation according to the spherical-drop model of Tables XXI and XXII will take place only if the initial radius is less than 0.052 (stone) or 0.054 cm (iron). For greater radii, there will be flattening or fissure, and the path of ablation will be shorter, about as given in Table XXVII.

CHAPTER 7

Atomic Collisions

7.1 Atomic Collisions as a Source of Ionization and Radiation

Meteors are rendered observable by radiation which they emit, by radio waves reflected from their wake trains, or by direct impact on receivers sent up in rockets.

Radiation emitted by meteors originates chiefly from excitation of the vapor atoms in collisions with the air molecules, or between themselves. This is borne out by meteor spectra which are line spectra of emission. Except in very slow meteors, visible continuous thermal radiation from the meteoroid does not exceed 1 per cent of the total.

The source of ionization is the same as that of radiation: collisions between heavy atomic particles without the participation of electrons. Electrons are not available in the beginning, and when a few are set free, their kinetic energies are far below those of the atoms; to attain equipartition of energy, electrons would have to undergo hundreds of collisions, yet time is too short for that. For this reason, the seat of meteoric excitation and ionization—the coma—is in a state of "underexcitation" as electron energy is concerned. Excitation and ionization there depend almost entirely on collisions between atoms which take place during the process of deceleration of the vapor atoms from meteor velocities to thermal velocities. Conditions of thermodynamic equilibrium are nonexistent, and the notion of temperature has no meaning.

The preceding strictly refers to the smaller meteors, chiefly those of Problem 1, including dustballs of all sizes. In massive meteors with an air cap, conditions may be different, and some approach to thermodynamic equilibrium may take place.

Experimental data with respect to atomic collisions are rather scanty, whereas absolute theory, capable of producing or predicting numerical results, is nonexistent. In view of this, a semi-empirical, semi-theoretical approach was undertaken by the writer in 1933,[7] on the basis of a combination of classical and quantum-mechanical principles; the theory represented observational data satisfactorily, with a logarithmic probable error of $\pm$ 0.19. Since then, new experimental data have not produced contradictory cases; some advance in theory has been achieved, as far as understanding of the mechanism is concerned,[47] without however yielding a clue to the numerical solution. Under such circumstances it seems that a restatement of the original theory, with some new supplementary calculations, is the best one can do for the present. There is nothing in existing visual, photographic, or radar observations which would suggest that the provisional theory of the efficiency of ionization and excitation is not correct to a close order of magnitude.

The theory[7] is principally one of ionization in atomic collisions. Excitation and ensuing radiation are calculated by assuming the same probability functions as for ionization. Whereas the calculated efficiencies of ionization were checked by experiment, those of excitation were not.

Failure to realize this led to a widespread misunderstanding, started by Herlofson[48] who, working backwards, estimated ionization efficiency from Öpik's radiation efficiency and applied a second time a correction factor of 10^{-2} which was already included in the luminous efficiencies; he obtained thus an efficiency of ionization which was too small by two orders of magnitude and arrived at the notorious ratio of energy dissipation divided "on heat, light and ionization in the ratios $10^4 : 10^2 : 1$."[48] These ratios are not only incorrect and disagree with experiment,[49] but also are unjustly ascribed to Öpik's theory. It is therefore regrettable that Herlofson's unfounded ratios are continually repeated and used by workers in the field. Lovell[50] persisted in ascribing the ratio to Öpik after being told by Öpik that the figures disagree completely with his results.[51]

The experimental figure for the line density of electrons in the wake train of 3rd mag Perseid meteors ($w = 60$ km/sec), as estimated by Greenhow and Hawkins[49] from Millman and McKinley's direct observational data,[52] is 9×10^{12} cm^{-1}. The calculated line density, according to Öpik's theory, is 7×10^{13} (all ions) or 1.4×10^{13} cm^{-1} (metallic ions only). Considering that the molecular ions are quickly eliminated by dissociative recombination,[9,26] the agreement of the predicted number of metallic ions with the observed number is quite satisfactory.

7.2 Potential Energy of Atomic Interaction

Atomic systems consist of nuclei with bound electrons. Although the masses of the electrons are small, they resist the penetration of another system to the limits of their binding energies. The Coulomb repulsion between the electrons and the Pauli exclusion principle play here a decisive role. Two colliding atoms form a temporary molecule, mostly with a negative binding energy (repulsion). In the case of chemical affinity and favorable (antiparallel) spin of the outer electrons, attraction and a weak positive binding energy (of a few ev) may result at moderate distances of approach, which at closer approach is replaced by normal repulsion. In the following, only a potential of repulsion is considered, monotonously increasing with decreasing distance. For the order-of-magnitude calculations the details of the interaction function are irrelevant (Fig. 6.).

Define an equivalent radius of the orbital of the nth electron as counted from outside:

$$r_n = 7.15 \times 10^{-8} n/I_n \text{ (cm)}, \tag{7-1}$$

where I_n is the ionization energy of the electron in ev. The type of orbital ($s, p, d, \ldots$) and screening are not taken into account.

Arrange the ionization energies I_n and I'_k and the corresponding radii r_n and r'_k of the two colliding atoms in the order of increasing energy, and let $I_n \leqq I'_k \leqq I_{n+1}$. If the energy of interaction at a distance $x_{nk} = r_n + r'_k$ is V_{nk}, the energy at a distance $x_{(n+1)k} = r_{n+1} + r'_k$ is assumed to be given by

$$V_{(n+1)k} = V_{nk} + I'_k + 14.3 \times 10^{-8} n(k-1)\left[\frac{1}{x_{(n+1)k}} - \frac{1}{x_{nk}}\right], \quad (7\text{–}2)$$

in electron-volts. The last term in Eq. 7–2 is supposed to represent Coulomb interaction.[7] The potential function can thus be built up step by step, according to Eq. 7–2. At distances greater than $r_1 + r'_1$, i.e., outside the orbitals, $V_0 = 0.036$ ev (thermal energy at 273° K) at $x_0 = R_0 + R'_0$ can be assumed, where R_0 and R'_0 are the gaskinetic radii of the two atoms or molecules.

TABLE XLI

Assumed Energies of Interaction

				a. Fe → N					
x, 10^{-8} cm	2.70	1.414	1.368	1.188	1.121	0.900	0.758	0.469	0.373
$V(x)$, ev	0.036	7.8	38.4	67.9	141	258	526	655	1025
				b. Mg → N					
x	1.452	1.440	0.743	0.714	0.647	0.639	0.350	0.343	0.311
$V(x)$	22.0	37.0	66.5	113.6	187.1	367	473	609	795
				c. Si → N					
x	1.379	1.364	1.128	1.096	0.656	0.589	0.569	0.289	0.283
$V(x)$	8.1	38.8	68.3	147	194	268	365	749	996
				d. O → N					
x	1.024	0.905	0.893	0.864	0.781	0.740	0.685	0.396	0.151
$V(x)$	13.6	28.0	57.5	92.5	194.3	345	442	555	1091
				e. N → N					
x	0.994	0.982	0.970	0.941	0.845	0.758	0.738	0.449	0.160
$V(x)$	14.4	28.8	58.3	87.8	182.0	329.0	426	523	1059

The energy of interaction is thus defined by a number of discrete values, $V = V(x)$. Interpolation between two discrete points V_1, V_2 may be conveniently made according to

$$V \sim x^{-s}. \tag{7–3}$$

As to diatomic molecules, at meteor velocities the molecular bond is relatively weak as compared with the kinetic energy, and interaction can be assumed to take place with one of the components of the diatom.

At meteor energies the Coulomb term in Eq. 7–2 does not exceed 5 per cent of the total and can be neglected. Table XLI contains discrete values of the interaction energy for meteor atoms with *atomic* nitrogen, calculated according to Eq. 7–2 with omission of the Coulomb term; x = distance between nuclei, $V(x)$ = interaction energy.

7.3 Elastic Energy Transfer

The inelastic fraction of energy (excluding dissociation) dissipated at meteor velocities is less than 1 per cent of the initial kinetic energy. In a first approximation, therefore, elastic dissipation can be assumed in calculating the range and deceleration of the meteor atoms. Collisions leading to dissociation are here also classified as elastic, involving only one component of a diatom. Disregarding thermal velocities, normally the system of reference will be fixed with respect to the free atmosphere, or the air molecule. In certain cases it may be referred to the center of mass of the cloud of vapors (coma).

The notations are those of Section 4.5; thus, μ = mass of moving atom, μ_1 = that of the atom of the medium, etc. Further x = distance between atoms, x_0 = periastron distance, or that of closest approach; $V = V(x)$ = energy of interaction (ergs or ev; repulsion positive, attraction negative); $V(\infty) = 0$; $V(x_0) = V_0$ = energy of interaction at periastron; ΔE = loss of kinetic energy by μ; ΔE_1 = gain of kinetic energy by μ_1; i = ultimate direction angle of motion of $\underset{1}{\mu}$, θ—of μ after the collision, both angles reckoned in opposite directions relative to the original direction of motion of μ; τ = target radius of the collision, or the distance of one of the atoms from the undeflected rectilinear path (asymptote) of the other (Fig. 13.).

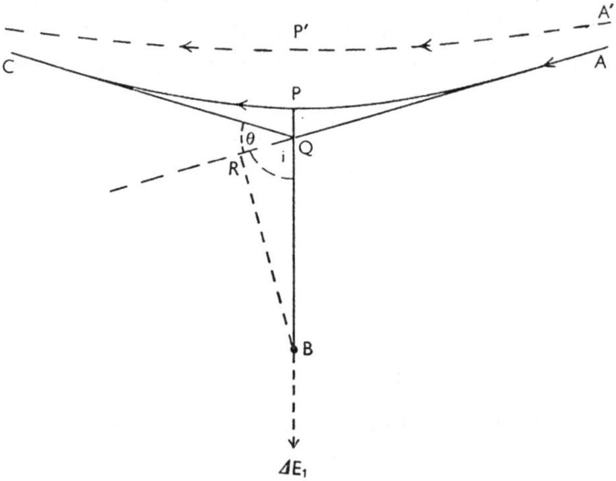

Fig. 13. Atomic encounter. Atom A (of mass μ) moves along APC, passing atom B (of mass μ). P is the periastron, BP = x_0 the periastron distance, AQ and QC the asymptotes, RQB = i, BR = τ, the target radius. τ = QB sin $i < x_0$ sin i. A'P' is the trajectory in a Coulomb field ($s = 1$), APC—in a field of high-power law of interaction ($s \cong 4$). The difference $x_0 \sin i - \tau$ = PQ sin i decreases with increasing s (power of the interaction law). B acquires kinetic energy ΔE_1 away from the periastron direction.

Irrespective of the form of $V(x)$, in addition to Eqs. 4–50, 4–51, and 4–54, for elastic collisions $\Delta E = - \Delta E_1$, and

$$\Delta E = - 4mE_r \cos^2 i/(1 + m), \tag{7–4}$$

$$V_0 = E_r[1 - (\tau^2/x_0^2)]. \tag{7–5}$$

For symmetry reasons, the ultimate momentum of $\underset{1}{\mu}$ is directed along the line of closest approach, and i is thus also the angle between the asymptote of the orbit and the direction of periastron.

τ/x_0 and V_0 depend upon i in a manner prescribed by the law of interaction (Fig. 13). With (7–3) as the law, $\tau/x_0 \sin i < 1$, and the ratio increases and approaches 1 with increasing s. Atomic interaction varies rapidly with distance, implying a high value of s. As an approximation we introduce the model of a rigid elastic sphere, with $s \to \infty$, but of a variable radius x_0, in each case corresponding to the energy of interaction V_0; the motion is then rectilinear until periastron, where the collision takes place with the full "accumulated" energy V_0. The relative trajectory consists in this case of two rectilinear portions, and

$$\tau = x_0 \sin i; \tag{7–6}$$

with this, Eqs. 7–5 and 7–4 lead to Eqs. 4–49 and 4–51. The simplification amounts to slightly overestimating the target radius.

Equation 7–4 (general) or Eq. 4–51 (approximate) define the elastic dissipation of kinetic energy, with the target radius being given by Eq. 7–5 (general), or by Eq. 7–6 (approximate).

The dissipation of kinetic energy takes place by finite accidental amounts, simultaneously with deflections (θ) in angle. The atom's track is a zig-zag line, a kind of random walk, consisting of unequal stretches and deflections. If Λ is the length of the zig-zag trajectory, Λ' the distance between the end points, the ratio $\Lambda'/\Lambda < 1$ depends upon the mass ratio and number of collisions. Consider the *average* energy dissipation along Λ which, according to Eq. 4–51, is

$$\frac{dE}{d\Lambda} = -\frac{4m\pi N}{(1+m)}\int_0^\infty V_0\,d(\tau^2), \tag{7–7}$$

where N is the number density of the atoms of the medium. For given E_r, or velocity, Eq. 7–5 and Table XLI define τ and V_0 for any value of x_0, and the numerical integration of Eq. 7–7 can be performed.

In Table XLII results of calculations for nitrogen-iron collisions are given. Y is the fraction of energy dissipated *at constant velocity* within subsequent limits of V_0, or τ. Instead of ∞, the upper limit of integration of Eq. 7–7 was set at $R^2 = 7.3 \times 10^{-16}$ cm² which is the square of the assumed gaskinetic target radius at near zero energy.

TABLE XLII

Target Radius (τ^2 in 10^{-16} cm²) and Relative Distribution of Elastic Energy Dissipation (Y) for Collisions of Nitrogen with Iron Atoms

($N \rightarrow Fe$, $m = 0.25$ or $Fe \rightarrow N$, $m = 4$) w = velocity, km/sec; E_r = relative energy, V_0 = interaction energy at periastron, both in electron-volts.

$w =$	14.8		20.9		29.6		41.8		59.2		83.6	
$E_r =$	12.8		25.6		51.2		102.5		205		410	
V_0	τ^2	Y	τ^2	Y	τ^2	Y	τ^2	Y	τ^2	Y	τ^2	Y
0	7.30	0.064	7.30	0.036	7.30	0.020	7.30	0.011	7.30	0.0061	7.30	0.0040
0.8	3.60	0.069	3.72	0.037	3.78	0.020	3.81	0.011	3.82	0.0064	3.83	0.0039
1.6	2.59	0.110	2.78	0.056	2.87	0.029	2.92	0.015	2.94	0.0085	2.95	0.0051
3.2	1.78	0.206	2.07	0.085	2.22	0.038	2.30	0.019	2.33	0.0100	2.35	0.0055
6.4	1.02	0.552	1.53	0.172	1.79	0.056	1.91	0.022	1.97	0.0072	2.03	0.0045
12.8	0.00	—	0.98	0.614	1.47	0.190	1.68	0.057	1.84	0.024	1.90	0.012
25.6	—	—	0.00	—	0.925	0.650	1.38	0.192	1.62	0.066	1.73	0.025
51.2	—	—	—	—	0.000	—	0.885	0.677	1.32	0.292	1.55	0.152
102.5	—	—	—	—	—	—	0.000	—	0.660	0.581	0.989	0.287
205	—	—	—	—	—	—	—	—	0.000	—	0.459	0.500
410	—		—		—		—		—		0.000	

For an elastic sphere of constant radius $Y_s = [(V_0)_1^2 - (V_0)_2^2]/E_r^2$, which yields the following "spherical" energy distribution:

V_0/E_r, limits 1	$\frac{1}{2}$	$\frac{1}{4}$	$\frac{1}{8}$	$\frac{1}{16}$	$\frac{1}{32}$	0
Y_s	0.7500	0.1875	0.0469	0.0117	0.0029	0.0010

As compared with this, the distributions in Table XLII are more widespread, although still markedly concentrated toward high energy intervals.

The half-energy range, over which the kinetic energy of μ is halved, is

$$\lambda = 0.69E/(-\,dE/d\Lambda), \tag{7–8}$$

where $0.69 = \ln 2$. λ is chiefly a function of velocity and mass ratio.

In Table XLIII a comparison is made between experiments by Durbin,[53] and data calculated according to Eqs. 7–8 and 7–7 for iron atoms in nitrogen.[7] The table is arranged in the order of increasing mass ratio, m. The calculated values fit satisfactorily in between those observed when allowance is made for the mass ratio.

TABLE XLIII

Ratio λ/λ_0 of Half-Energy Range to Gaskinetic Free Path[53]

Pair	m		E_r, ev 3	10	30	100
$K^+ \rightarrow A$	1.0	Observed $\lambda/\lambda_0 =$	1.5	2.5	3.8	5.0
$K^+ \rightarrow N_2(N)$	1.4 (2.8)	Observed $\lambda/\lambda_0 =$	2.3	3.0	3.6	4.2
$Fe \rightarrow N$	4	Calcul. $\lambda/\lambda_0 =$	3.1	4.1	5.4	6.4
$K^+ \rightarrow He$	10	Observed $\lambda/\lambda_0 =$	4.9	7.6	12	—

When the roles of moving atom and medium are interchanged, or when $m_1 = 1/m$, $E_1 = E/m$, and $N_1 = N$, according to Eqs. 7–8 and 7–7, $\lambda_1 = \lambda$, or the relative rate of dissipation is the same, whereas the absolute rate is proportional to the atomic weight. Fe atoms will be stopped by the same number of collisions when moving in nitrogen, as will N atoms moving in iron vapors. However, the Fe atoms in a nitrogen medium will be less deflected in angle than the N atoms in an iron medium.

7.4 Total Path and Transversal Penetration of Atoms

The interpretation of radar observations requires a knowledge of the radius of the ionized column. This radius increases from gas and eddy diffusion, a phenomenon related to the fading of radar echoes. Actually, gas diffusion alone has been taken into account in various discussions of the subject,[48, 50, 54, 55, 56] an oversimplification which is not rendering justice to the realities of the problem. Although for faint

TABLE XLIV

Total Range (λ) and Transversal Range (R) for Single Atoms Moving in Nitrogen

w = initial velocity, km/sec; E_r = initial relative kinetic energy in electron-volts; $\rho\lambda$ and ρR, total and transversal air masses in 10^{-8} g/cm². λ, R are half-energy ranges; $\lambda(\frac{1}{4})$, $R(\frac{1}{4})$—quarter-energy ranges; $\lambda(0)$, $R(0)$—ranges to room-temperature energies

	E_r	820	410	205	102.5	51.2	25.6	12.8	6.4
Atom	w	118.4	83.6	59.2	41.8	29.6	20.9	14.8	10.4
	$\rho\lambda$	13.7	10.9	9.3	8.2	7.6	7.2	4.7	4.5
	$\rho\lambda(\frac{1}{4})$	24.6	20.2	17.5	15.8	14.8	11.9	9.2	8.3
Fe → N	$\rho\lambda(0)$	82.6	68.9	58.0	48.7	40.5	32.9	25.7	21.0
	ρR	3.6	2.7	2.2	1.8	1.8	1.8	1.1	1.0
	$\rho R(\frac{1}{4})$	8.8	6.7	5.7	5.2	4.9	3.9	3.1	2.7
	$\rho R(0)$	56.0	45.0	36.8	30.3	24.7	19.0	15.0	11.8
	w	133.2	94.0	66.6	47.0	33.3	23.5	16.6	11.7
	$\rho\lambda$	31.5	25.3	21.8	16.1	8.2	6.8	4.4	4.3
	$\rho\lambda(\frac{1}{4})$	56.8	47.1	37.9	24.3	15.0	11.2	8.7	7.8
Mg → N	$\rho\lambda(0)$	134.0	102.5	77.2	55.4	39.3	31.1	24.3	19.9
	ρR	9.6	7.4	6.1	4.4	2.3	1.9	1.3	1.2
	$\rho R(\frac{1}{4})$	23.1	18.5	14.1	8.4	5.6	4.1	3.4	2.8
	$\rho R(0)$	57.3	41.8	31.8	22.2	17.4	14.1	11.4	9.1
	w	129.6	91.5	64.8	45.8	32.4	22.9	16.2	11.4
	$\rho\lambda$	35.3	26.7	14.9	8.8	7.8	7.5	5.0	4.9
	$\rho\lambda(\frac{1}{4})$	62.0	41.6	23.7	16.6	15.3	12.5	9.9	9.0
Si → N	$\rho\lambda(0)$	129.0	93.7	67.0	52.1	43.3	35.5	28.0	23.0
	ρR	10.1	7.4	4.0	2.2	2.2	2.1	1.3	1.3
	$\rho R(\frac{1}{4})$	23.8	14.6	8.1	5.8	5.7	4.5	3.5	3.1
	$\rho R(0)$	57.8	41.2	30.6	24.5	20.9	17.3	13.7	12.1
	w	145.0	102.4	72.5	51.2	36.2	25.6	18.1	12.8
	$\rho\lambda$	27.7	15.6	15.1	15.8	16.3	14.8	8.0	6.7
	$\rho\lambda(\frac{1}{4})$	43.3	30.7	30.9	32.1	31.1	22.8	14.7	12.4
O → N	$\rho\lambda(0)$	144.7	117.0	101.4	86.3	70.5	54.2	39.4	31.4
	ρR	9.4	5.3	4.9	5.4	5.7	5.0	2.7	2.3
	$\rho R(\frac{1}{4})$	18.6	14.2	14.5	18.2	14.3	9.7	6.7	5.7
	$\rho R(0)$	48.1	40.6	35.5	30.9	24.7	18.7	14.3	11.8
	w	149.8	105.9	74.9	52.9	37.5	26.4	18.7	13.2
	$\rho\lambda$	25.1	13.9	13.6	13.8	14.2	14.2	8.9	8.4
	$\rho\lambda(\frac{1}{4})$	39.0	27.5	27.4	28.0	28.4	23.1	17.3	15.4
N → N	$\rho\lambda(0)$	144.1	119.0	105.1	91.5	77.7	63.5	49.3	40.4
	ρR	10.3	5.7	5.6	5.6	5.8	5.8	3.6	3.4
	$\rho R(\frac{1}{4})$	19.8	14.9	14.9	15.2	15.4	11.9	9.3	8.0
	$\rho R(0)$	38.9	33.5	30.9	26.9	23.5	19.0	15.1	12.7

meteors turbulent (eddy) diffusion is unimportant, the initial penetration of the meteor atoms and ions during the process of deceleration to thermal velocities is of decisive importance and its neglect has rendered some of the results illusory. The half-energy range, itself a multiple of the gaskinetic free path, is not yet a measure of the penetration. Before thermal velocities are attained, the atom passes through a succession of half-energy ranges, and its total range is very much greater. With $\lambda_0\rho = 10^{-8}$ g/cm² for the gaskinetic free path (λ_0), from Table XXIII (which has been calculated from Eq. 7–8), it can be seen that at 40 km/sec a nitrogen atom has a half-energy range $\lambda \cong 14\lambda_0$ in air, and a total range along a zig-zag path of about $80\lambda_0$. The transversal penetration will be smaller, but still a large multiple of λ_0.

The theory of transversal spread is a complicated problem of random walk (Öpik, to be published). The problem is essentially one of anisotropic diffusion whose rate, however, is very much greater than that of ordinary gas diffusion. Some results are contained in Table XLIV. In the table are given the products $\rho\lambda$ and ρR, in 10^{-8} g/cm² units, where R is the transversal penetration. To obtain the linear range in cm, the tabular values must be multiplied by $10^{-8}/\rho$; λ is the half-energy range, $\lambda(\frac{1}{4})$—the quarter-energy range, $\lambda(0)$—the total zig-zag length of path until the kinetic energy is reduced to thermal energies, or the cumulative sum of the half-energy ranges. With $\rho\lambda_0 = 10^{-8}$ for the gaskinetic free path, the figures as printed in Table XLIV are practically expressed in units of λ_0. The same is true of the transversal range, R, except that the total range, $R(0)$, is not the cumulative sum of R; R, $R(\frac{1}{4})$ and $R(0)$ are mean square deviations at right angles to the initial rectilinear path.

7.5 Dilution of Coma

Table XLIV is calculated for single atoms shooting through the medium, i.e., for a great dilution of the meteor substance in air; let us investigate the degree of dilution.

If ν_i is the line density of a given kind i of atoms, i.e., the number of atoms released (vaporized) per cm of the path, the number density of these atoms per cm³ at a distance y from the axis of the wake train is given by

$$N_i = (\nu_i/\pi R_i^2) \exp(-y^2/R_i^2), \tag{7-9}$$

where R_i is the mean square transversal range depending on the phase of dissipation (Table XLIV) (Fig. 14.).

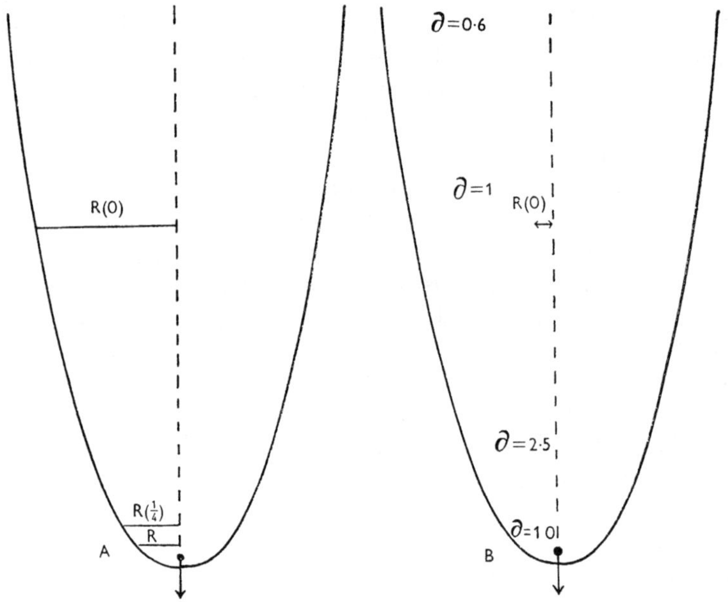

Fig. 14. Development of coma. A, small meteoroid and greatly diluted coma; envelope of coma defined by atomic penetration, R (at half-energy phase), $R(\frac{1}{4})$ (at quarter-energy phase), and $R(0)$ (at thermal energy phase). B, large meteoroid and dense cloud of vapors; envelope of coma defined by the expanding vapor front; atomic penetration, R or $R(0)$, is small as compared to coma radius; dilution factor (∂) inversely proportional to coma cross section.

The material density of the dissipated atoms at a distance y is

$$\rho_m = \sum_i \mu N_i. \tag{7-10}$$

Let us define a dilution factor as the ratio of meteor vapor density to air density, $\partial = \rho_m/\rho$. If L is the effective length of the meteor trail (path of ablation), M the mass, $M = \Sigma M_i$,

$$\nu_i \underset{i}{\mu} \cong M_i/L. \qquad (7\text{–}11)$$

On the axis of the coma and wake train, $y = 0$, $\underset{i}{\mu} N_i = \underset{i}{\mu} \nu_i/(\pi R_i^2)$ according to Eq. 7–9 and, from Eq. 7–10, with the line mass density according to Eq. 7–11, the dilution ratio on the coma axis becomes

$$\partial_0 = \Sigma[M_i/(\pi R_i^2 L\rho)]. \qquad (7\text{–}12)$$

Setting $R_i = 10^{-8} z_i/\rho$, so that z_i is numerically equal to the tabular value of ρR as printed in Table XLIV, and with $\vartheta_i = M_i/M$, we obtain

$$\partial_0 = 10^{16}\rho M\Sigma(\vartheta_i/z_i^2)/(\pi L). \qquad (7\text{–}13)$$

For stony dustballs an initial dispersion of z_0^2, due to the spread of the dustgrains, must be added to z^2; further, with $L = 1.5(H_a - H_d) \simeq 2 \times 10^6$ cm (Table XVIII), $\rho \simeq 2 \times 10^4/w^2$, (7–13) becomes

$$\partial_0 = 3 \times 10^{13} M/[w^2(\bar{z}^2 + z_0^2)], \qquad (7\text{–}14)$$

where $\bar{z}^2$ is defined as a harmonic mean,

$$\bar{z}^2 = 1/\Sigma(\vartheta_i/z_i^2). \qquad (7\text{–}15)$$

The linear dispersion of the dustgrains at Middle Trail, or $10^6/w$ seconds after the flashpoint, is about $10^8/w$ cm, if their radial velocity at breakup is about 100 cm/sec (Section 4.2). In units of λ_0, or those of z, this is $10^{16}\rho/w$ and, with ρ as for dustballs, the linear spread is $2 \times 10^{20}/w^3$. Hence $z_0^2 = 4 \times 10^{40}/w^6$. With this, Table XLV is calculated from (7–14) and (7–15), assuming a schematic composition of meteoric stone as follows:

Element	Fe	Mg	Si	O
Fraction by weight, ϑ_i	0.15	0.17	0.23	0.45

TABLE XLV

Ratio of Dilution Factor on Coma Axis to Mass for Average Dustball Coma

$\bar{E}_r$, ev	410	205	102.5	51.2	25.6	12.8	6.4
w, km/sec	95.6	67.7	47.8	33.9	23.9	16.9	12.0
z_0^2	0.05	0.42	3.4	26.4	216	1720	13400
	At half-energy phase, R						
$\bar{z}^2$	22.8	14.5	8.5	6.9	6.3	2.37	2.04
∂_0/M	0.014	0.044	0.110	0.079	0.024	0.0061	0.0016
	At quarter-energy phase, $R(\frac{1}{4})$						
$\bar{z}^2$	140	90	62	38	27.7	16.9	12.5
∂_0/M	0.0023	0.0073	0.020	0.032	0.022	0.0060	0.0016
	At thermal energy phase, $R(0)$						
$\bar{z}^2$	1720	1130	730	480	300	185	126
∂_0/M	0.00019	0.00058	0.0018	0.0052	0.010	0.0055	0.0016

Thus, for $M = 0.1$ g and $w = 23.9$ km/sec, which corresponds to a visual meteor of about the 3rd magnitude (absolute), the dilution factor, ∂_0, is 0.0024, 0.0022, and 0.0010 for R, $R(\frac{1}{4})$, and $R(0)$, respectively. These factors are very small and, therefore, the figures of Table XLIV would apply in this case. For M as large as 3 g (absolute magnitude -1.5) ∂_0 will be still small enough for the single-atom model of Table XLIV to apply.

For compact meteoroids the general Eq. 7–13 applies. Substituting M from Eq. 4–1 and ρ from Eq. 5–36 as for *normal environment*, with $\cos Z = \frac{2}{3}$, the dilution factor on the coma axis becomes

$$\partial_0 = 8.9 \times 10^{15} \phi r_0^4 \delta/(w^2 L \bar{z}^2). \qquad (7\text{–}16)$$

With $\phi = \phi_c$ as for middle trail (Table XXII), $L = 2.7 \times 10^6$ cm for Cases 4 and 7 which only are of interest in this context (moderately large meteors), the following relations are obtained:

$$\partial_0 = 8.1 \times 10^{15} r_0^4/(w^2 z^2) \text{ (iron, Case 4)}, \qquad (7\text{–}17)$$

where z is to be taken directly from the upper section of

Table XLIV (Fe → N); and

$$\partial_0 = 7.9 \times 10^{15} r_0^4/(w^2 \bar{z}^2) \text{ (stone, Case 7).} \quad (7\text{–}18)$$

Thus, for stone, with $r_0 = 0.5$ cm, $w = 33.9$ km/sec, and for the ultimate phase $R(0)$ of dissipation, $\bar{z}^2 = 480$, $\partial_0 = 0.090$.

7.6 Inelastic Energy Transfer and Dissociation of Diatoms

The maximum inelastic energy which can be absorbed in an atomic collision is equal to V_0; if this takes place, the energy transfer formulae in a "perfectly inelastic" collision, instead of Eq. 7–4, become, together with Eq. 4–49:

$$\Delta E = -V_0[1+m/(1+m)], \quad \Delta E_1 = +V_0 \cdot m/(1+m). \quad (7\text{–}19)$$

An actual case may be intermediate between Eqs. 7–4 and 7–19.

The maximum ratio of inelastic energy to the total dissipated is thus

$$\Delta E_i/\Delta E = (1+m)/(1+2m). \quad (7\text{–}20)$$

In a perfectly inelastic collision of an atom μ with one component of a diatom $2\underset{1}{\mu}$, $m = \mu/\underset{1}{\mu}$ as before, simultaneously with the creation of inelastic energy V_0, the amount ΔE_1 Eq. 7–19 is transferred as kinetic energy to $\underset{1}{\mu}$; according to Eq. 4–50 with $m = 1$, $\frac{1}{2}\Delta E_1$ is available for internal vibrational energy; and, to remain a diatom, this must be less than the energy of dissociation, i, whence the maximum amount of inelastic energy which can be created without breaking up the diatom becomes

$$V_0 \leqq 2i(1+m)/m. \quad (7\text{–}21)$$

On the other hand, dissociation without ionization or excitation in a nominally elastic collision can take place more easily, because in this case the kinetic energy transfer is four times that for a perfectly inelastic collision. To cause dissociation, the transfer Eq. 4–51 must be at least double the dissociation energy, $\Delta E \geqq 2i$, or

$$V_0 \geqq i(1 + m)/2m. \tag{7-22}$$

Table XLVI contains the limits, calculated from Eqs. 7–21 and 7–22.

TABLE XLVI

Maximum Inelastic Energy Without Dissociation, V_0 (max)
Minimum Energy of Interaction for Dissociation, V_0 (min)

Hitting atom	Fe	Mg	Si	O	N
		N_2; $i = 9.76$ ev			
m	4.00	1.71	2.00	1.14	1
V_0 (max), ev	24.4	31.0	29.3	36.6	39.0
V_0 (min), ev	6.1	7.8	7.3	9.2	9.8
		O_2; $i = 5.08$ ev			
m	3.50	1.50	1.75	1.00	0.875
V_0 (max), ev	13.1	17.0	16.0	20.3	21.8
V_0 (min), ev	3.3	4.2	4.0	5.1	5.4

From Table XLVI it appears that ionization of atmospheric molecules, N_2 (15.6 ev) and O_2 (12.2 ev), is possible through impact of meteor and air atoms. However, the range of V_0 at which molecular ionization can be produced is narrow and the probability low (Table XLII). Atmospheric molecules are more likely to become ionized through exchange of charge. On the contrary, $V > V_0$ (min) covers over 90 per cent of the energy transfer (Table XLII), and dissociation is readily achieved. The efficiency of dissociation is so great that molecules will be broken up practically in each collision, whereas ionization and excitation of the molecules will occur in, say, only one collision out of one thousand.

7.7 Probability of Ionization and Excitation

Let the total quantal energy of the two colliding atoms be

$$\Omega_i = \Sigma I_n + \Sigma I'_k + \varepsilon_n + \varepsilon'_k, \tag{7-23}$$

up to a certain level (n, k) of ionization (Section 7.2, $i = n + k$); ε_n and ε'_k denote excitations of the ionized states. The necessary condition for an inelastic transfer (Eq. 7–23) to take place is

$$V_0 \geqq \Omega_i\,. \tag{7–24}$$

The condition, however, is not sufficient. Most collisions are elastic and, despite condition 7–24 being fulfilled, quantal changes will not occur. In the few cases when quantal changes are produced, different combinations of ionization and excitation will satisfy Eq. 7–24.

For practical reasons a further simplification is introduced; it is assumed that *when* the collision is inelastic, the maximum possible quantal energy Ω_i compatible with Eq. 7–24 is transferred; or, that the nearest to a perfectly inelastic collision takes place in such a case. The incorrectness of this assumption is automatically compensated by an empirically adjusted coefficient of efficiency.

According to Eq. 7–23, the possible values of inelastic energy form a sequence of increasing discrete amounts. Let Ω_i and Ω'_i be two such values, $\Omega'_i > \Omega_i$. The target radius τ_i corresponding to $V_0 = \Omega_i$ is defined by Eq. 7–5 or Eq. 7–6, and similarly the target radius τ'_i corresponding to $V'_0 = \Omega'_i$. The cross section over which the inelastic transfer Ω_i is *possible* (without necessarily taking place) is thus

$$\sigma_i = \pi[(\tau'_i)^2 - \tau_i^{\,2}]. \tag{7–25}$$

The cross section for the inelastic transition is then

$$\sigma(\Omega_i) = k_i \sigma_i, \tag{7–26}$$

where k_i is the probability of the inelastic transition. From experimental data[57, 58] on ionization by ionic beams of widely differing energy, including the range of meteor velocities, it was found[7] that the probability could be represented by an empirical expression,

$$k_i = 0.026(E_r/\Omega_i)^{\frac{1}{2}}, \tag{7–27}$$

with the double restriction $E_r \geqq \Omega_i$ and $k_i \leqq 1$.

In interpreting experimental data, it was assumed that ions of higher order are reduced to singly-charged ions by exchange of charge. Excluding alpha-particles, for which exact agreement was obtained, the rest of the data yielded a logarithmic probable error of $\pm$ 0.19 in the ratio of observed to computed number of ions. The result is very satisfactory and favorably compares with the accuracy of quantum-mechanical calculations of transition probabilities.

7.8 Ion Distribution in the Coma

The results of this section, as those of Table XLIV, refer to a diluted coma, $\partial_0 \ll 1$. Only the ultimate phase $R(0)$ of the coma, corresponding to the "initial spread" of the ionized column, is considered (Fig. 14.). This phase marks the beginning of the spread of the wake train by gas and turbulent diffusion. On principles exposed in the preceding sections, the ionization and spread of ions has been calculated by numerical integrations. Primary collisions of the meteor atoms with air molecules, as well as secondary and higher order collisions of the atmospheric ions and their further spread during the process of dissipation have been taken into account.

In a diluted coma, secondary collisions are taking place mainly between the atmospheric constituents. The atmospheric ions, as well as doubly and multiply ionized metal atoms are assumed to transfer their charge to the O_2^+ ions. The result is a nonhomogeneous ionized column, consisting of singly ionized metal atoms, some doubly ionized calcium atoms, and ionized oxygen molecules. Table XLVII contains the results of the calculations. In the section for iron, the third line gives the ultimate degree of ionization.

When the ion density is sufficiently high, the molecular ions will rapidly decay and metal ions alone will remain.[9]

When ∂_0 is large as for massive meteors, the conditions will differ from those of Table XLVII; the O_2^+ and higher order metal ions will more frequently meet neutral metal atoms and exchange charge with them; as a result, the metal

TABLE XLVII

Ionizing Efficiency and Spread of Ions in Air

w = initial velocity of atom, km/sec; n_1 = total number of singly ionized metal (Fe, Mg, Si) atoms, n_2 = total number of O_2^+ ions in 10^7 units per erg of initial kinetic energy, created in the deceleration from w to thermal velocities; $z = \rho R(0)$ (cf. Table XLIV) (z_1 for metal ions, z_2 for O_2^+ ions) is the transversal spread in units of 10^{-8} g/cm² (practically in units of the gaskinetic free path, λ_0); w_0, km/sec, is the lower limit of initial velocity for collisional ionization to take place

w	14.8	20.9	29.6	41.8	59.2	83.6	
			Iron				w_0
n_1	2.5	12.0	17.4	16.4	12.7	9.1	11.6
z_1	15.0	19.0	24.7	30.3	36.8	45.0	
$Fe^+/(Fe + Fe^+)$	0.0025	0.024	0.071	0.134	0.207	0.297	
n_2	0	1.0	9.2	23.5	39.8	56.8	17.0
z_2	—	11.4	12.9	14.1	15.2	16.3	
			Magnesium				w_0
n_1	1.4	10.9	20.7	22.0	17.8	13.0	13.0
z_1	10.5	13.1	15.9	20.1	28.6	38.2	
n_2	0	0.1	5.8	19.2	34.9	52.0	20.0
z_2	—	10.9	12.4	14.1	16.7	19.7	
			Silicon				w_0
n_1	1.8	11.2	20.2	20.7	16.4	12.1	12.7
z_1	10.7	16.4	20.1	23.6	28.7	38.0	
n_2	0	0.2	6.5	20.0	35.8	52.9	19.4
z_2	—	11.0	12.6	13.9	15.4	18.4	
			Oxygen				w_0
n_2	0	1.6	9.0	20.5	37.0	53.3	20.0
z_2	—	10.6	12.8	15.3	17.1	17.9	
			Stone, average				w_0
n_1	1.0	6.2	10.9	11.0	8.7	6.4	11.6
z_1	12.0	15.7	19.0	23.0	30.0	39.7	
n_2	0	0.9	7.9	20.6	36.8	53.5	17.0
z_2	—	10.7	12.8	14.5	16.2	18.0	

atoms will take over the entire charge, and only after they are practically all singly ionized, will O_2^+ and other higher ions survive. Also, in these more massive meteors, the spread of the ions will be impeded by Joule losses from electrons trailing behind; it is estimated that this happens for meteors brighter than -2 mag at $H = 84$ km, and those brighter than -4.5 mag at $H = 96$ km. Thus, the calculations are valid for most "dustball" cases, or for visual and telescopic meteors, but not for the brighter photographic fireballs. In these the initial spread is through expansion of the gaseous cloud of the coma.

CHAPTER 8

Meteor Radiation

8.1 Radiative Efficiency of Atomic Excited States

An excited state will either release its energy through radiation in cascade transitions to a lower level, ultimately to the ground state; or it will be de-excited by a collision of the "second kind" with another atom. The cross section for de-excitation being normally smaller than the gaskinetic cross section, the minimum time of unobstructed existence of an excited state is $t_0 \geqq \lambda_0/w$. Thus, at $H = 90$ km, $w = 20$ km/sec, $\lambda_0 = 2.1$ cm, $t \geqq 10^{-6}$ sec. The lifetime of most excited states is 10–100 times shorter, and they will have time to emit their radiation before being de-excited by a collision. Practically, therefore, excitation and radiation are synonymous in meteors. However, metastable or semi-metastable states, of a lifetime longer than 10^{-5}–10^{-4} sec, will be practically eliminated by collisions during the flight of the atom. Forbidden lines are therefore not expected to be observable in meteors except from metastable states at thermal velocities (Section 3.4).

When $G = 1/t_0$ is the collisional damping constant, or the number of de-exciting collisions per second, and A' the Einsteinian quantum-mechanical damping constant, or the inverse lifetime of an excited state, the damping (quenching) ratio or the factor by which the intensity of radiation is suppressed on account of collisions, is

$$q = A'/(A' + G). \tag{8-1}$$

For some excited states of meteor atoms the visual efficiency, β, for $G = 0$ (unobstructed cascade transitions)

TABLE XLVIII

Visual Radiative Efficiencies of Atomic Excited States

Atom	Fe	Fe	Fe	Fe	Fe	Fe	Fe	Fe	Fe
State	z^7D^0	z^7F^0	z^7P^0	z^5D^0	z^5F^0	z^5P^0	z^3F^0	z^3D^0	z^3P^0
ε, ev	2.39	2.80	2.93	3.20	3.32	3.59	3.86	3.87	4.19
A', sec^{-1}	5000	22000	16000	5.4×10^6	5.2×10^6	3×10^7	1.2×10^6	7×10^5	5×10^7
β $(G=0)$	0.690	0.190	0.048	0.220	0.100	0	0.370	0.342	0.001
β $(G=10^6)$	0.003	0.004	0.001	0.185	0.084	0	0.204	0.140	0.001

Atom	Fe	Fe	Fe	Fe	Fe	Fe+	Fe+	Fe+	Fe+
State	z^3G^0	e^7D	e^5D	e^5F	e^7F	z^6D^0	z^6F^0	z^6P^0	z^4F^0
ε	4.37	5.29	5.52	5.80	6.21	4.75	5.21	5.27	5.46
A'	6×10^7	4×10^8	7×10^8	4×10^9	2×10^8	—	—	—	—
β $(G=0)$	0.027	0.449	0.086	0.019	0.110	0	0.0029	0.0856	0.0354
β $(G=10^6)$	0.027	0.307	0.077	0.019	0.022	0	0.0029	0.0856	0.0354

Atom	Fe+	Fe+	Na	Na	Na	Na	Mg	Mg	Mg
State	z^4D^0	z^4P^0	3^2P^0	4^2S	3^2D	4^2D	3^3P^0	3^1P^0	4^1S
ε	5.49	5.80	2.09	3.18	3.60	4.27	2.70	4.33	5.09
A'	. . .	. . .	6×10^7	2.5×10^7	5×10^7	2×10^7	400	2×10^8	5×10^7
β $(G=0)$	0.0018	0	0.038	0.025	0.022	0.067	0.409	0	0.635
β $(G=10^6)$	0.0018	0	0.037	0.024	0.021	0.063	0.0002	0	0.406

Atom	Mg	Mg	Mg	Mg	Mg+	Mg+	Mg+	Mg+
State	4^1S	3^1D	4^3P	3^3D	3^2P–4^2D	4^2F^0	5^2P^0	6^2S
ε	5.37	5.73	5.91	5.92	4.41–11.52	11.58	12.03	12.74
A'	2×10^7	4×10^7	1.3×10^7	1.4×10^8	—	2.4×10^8	—	—
β $(G=0)$	0	0	0.546	0.187	0	0.0656	0.0001	0.0141
β $(G=10^6)$	0	0	0.318	0.0009	0	0.0654	0.0001	0.0140

Atom	O	O	O	O	N	N	N
State	3^5S^0–4^5P	4^3P	5^5S^0	6^5S^0	$3s^4P$–$3d^2D$	$4p^2S^0$	$4p^4D^0$
ε	9.11–12.23	12.31	12.61	12.96	10.28–12.98	13.14	13.18
β $(G=0)$	0	0.0046	0	0.0370	0	0.0413	0.0147

Atom	N	N	Si	Si	Si	Si	Si	Si+
State	$4p^4P^0$	$4p^4S^0$	$4s^3P^0$–$4d^3P^0$	$5p^3D$	$5p^3P$	$4d^3F^0$	$5p^3S$	—
ε	13.21	13.26	4.90–7.00	7.03	7.07	7.08	7.10	$\leqq 9.79$
β $(G=0)$	0.0026	0.0005	0.0002(aver.)	0.0027	0.0102	0.0002	0.0173	0

Atom	Si+	Si+	Si+	Si+	Ca	Ca	Ca	Ca	Ca+
State	4^2p^0	5^3S	4^2D	4^2F^0	—	4^1P^0	5^3S	5^1S	4^2P^0
ε	10.02	12.09	12.47	12.47	$\leqq 2.70$	2.92	3.89	4.11	3.11
β $(G=0)$	0.0006	0.0032	0.0580	0.0079	0	0.0354	0.0031	0.0251	0.0021

and $G = 10^6$ (order of magnitude of collisional damping in meteors) is given in Table XLVIII. The table is an excerpt from more detailed data.[26] ε is the excitation energy in electron-volts from the ground state. The visual radiative efficiency, β, is the fraction of ε which is converted into visible radiation. The downward cascade of transitions from the given excited level leads to the total energy ε being split up into amounts $i_1, i_2, \ldots, i_k$, radiated in certain discrete wavelengths. To each wave length corresponds a quenching ratio q_i, (cumulative over the cascade) and a certain sensitivity factor of the human eye, b_i. The visual efficiency of the excited level is then

$$\beta = \Sigma q_i b_i i_i / \varepsilon. \tag{8-2}$$

The sensitivity factors b_i were assumed according to Nutting[40] as summarized in Table XLIX. In view of the

TABLE XLIX
Visual Sensitivity Factor, b
(l = wavelength in Angstroms)

l	b
3800	0.001
4000	0.012
4200	0.048
4400	0.196
4600	0.450
4800	0.740
5000	0.990
5100	1.000
5200	0.860
5400	0.520
5600	0.240
5800	0.075
6000	0.018
6200	0.005
6400	0.001

Purkinje effect, the table should apply to meteors between the sixth and first magnitudes, with an optimum at 3.5 mag, and should be still usable at − 4 mag.

The distribution of energy in the downward cascade[26] was partly calculated from theoretical transition probabilities and with the aid of the tables of Bates and Damgaard[59] (for the lighter atoms); and partly from arc intensities of spectral lines calibrated by a few absolute determinations (for Fe).

The average visual efficiency of an atom can be defined as

$$\beta_a = \Sigma(\Pi_i \beta_i \varepsilon_i)/\Sigma(\Pi_i \varepsilon_i), \qquad (8\text{–}3)$$

where Π_i is the probability of excitation, and the summation is made over all discrete states. The probabilities of excitation in atomic collisions are unknown, but it appears that low levels are more easily excited than high levels. Following the empirical lead of Eq. 7–27, it was assumed that the probability is proportional to $\varepsilon^{-\frac{1}{2}}$ and to the interval $\Delta\varepsilon$ between adjacent excitation levels,

$$\Pi_i = \varepsilon^{-\frac{1}{2}} \Delta\varepsilon_i / \Sigma(\varepsilon_i^{-\frac{1}{2}} \Delta\varepsilon_i). \qquad (8\text{–}4)$$

TABLE L

Mean Visual Luminous Efficiency of Atoms, β_a

(conventionally weighted averages for atomic impact radiation; G = frequency per second of collisions of the second kind; I = ionization potential, volts)

Atom	I	G	β_a
Fe	7.87	100	0.0592
Fe	7.87	10^6	0.0367
Fe^+	16.18	100–10^5	0.0110
Na	5.14	0–10^5	0.0438
Na^+	47.3	0–10^5	0.00013
Mg	7.64	0	0.1640
Mg	7.64	3×10^5	0.0425
Mg	7.64	3×10^6	0.0347
Mg	7.64	3×10^7	0.0173
Mg^+	15.03	0–10^5	0.0063
O	13.61	100–10^5	0.0059
N	14.54	10–10^5	0.00073
Si	8.15	100–10^5	0.0041
Si^+	16.34	0–10^5	0.0060
Ca	6.11	0–10^4	0.0570
Ca^+	11.87	0–10^5	0.0055
H	13.59	0–10^5	0.0022

Instead of single levels, nearby levels were joined into groups of average ε and β, and Eqs. 8–4 and 8–3 were applied to these "combination levels." The results of the calculation[26] are collected in Table L.

The average visual radiation output, in electron-volts per excited atom, equals

$$Y = \Sigma \Pi_i \beta_i \varepsilon_i . \tag{8–5}$$

According to Eq. 8–4,

$$\Sigma \Pi_i \varepsilon_i = \int_0^I \varepsilon^{\frac{1}{2}} \, d\varepsilon \Big/ \int_0^I \varepsilon^{-\frac{1}{2}} \, d\varepsilon = \tfrac{1}{3} I, \tag{8–6}$$

where I is the ionization energy. Hence, substituting Eqs. 8–6 and 8–5 into Eq. 8–3, we obtain

$$Y = \tfrac{1}{3} I \beta_a . \tag{8–7}$$

8.2 Luminous Efficiency of Meteor Impact Radiation

The luminous efficiency of meteor vapors can be represented as the sum of expressions 8–7 for the different ionized states created in the process of dissipation of the kinetic energy of the vapor atoms. The excited states radiate their energy at once; whereas the resulting multiply ionized ground states transfer their charge to the medium and ultimately reduce to singly charged ions.

In the process of redistribution of charge the surplus energy may result in excitation. Thus, in $Fe^{++} + O \rightarrow Fe^{+} + O^{+}$ the surplus energy is 2.57 ev; only low-lying metastable states of Fe^{+} or O^{+} can be created which will be de-excited by collisions, and negligible radiation will result. Similar is the case of other exchange combinations with air molecules, which prevail in visual meteors with a diluted coma. Therefore, in the calculations of luminous efficiency, radiation caused by exchange of charge has been neglected. Some contribution from this source may be perceptible in bright meteors with a feebly diluted coma, when collisions between meteor atoms are frequent and when, e.g., in the exchange reaction $Fe^{++} + Fe \rightarrow Fe^{+} + Fe^{+}$ the surplus energy of 8.31 ev could raise Fe^+ to various normal (i.e., not metastable) excited levels (cf. Table XLVIII) and lead to some observable radiation. However, the neglect of this source is not of much consequence, because the radiative efficiency of ionized atoms (cf. Table L), especially that of the higher degrees of ionization, is low.

Let us first consider the case of a diluted coma, the meteor atom experiencing collisions with air molecules only. The luminous efficiency of the latter or their dissociated components is small and can be neglected (Table L). Visible radiation will be thus produced by the meteor atom only. In the course of its flight, a certain statistical equilibrium between the various ionized states as depending on velocity will be established; actually, the neutral and first ionized states will be predominant, the higher states having a very short life on account of exchange of charge.

Let P_n be the probable number of excitations of the nth ionized state, experienced by the atom while decelerating from initial velocity w to thermal velocity. The total luminous efficiency, or the fraction of initial kinetic energy converted into visible light, is then according to Eq. 8–7,

$$\beta = \tfrac{1}{3}\Sigma(P_n I_n \beta_n / E), \tag{8–8}$$

where E (ev) is the initial kinetic energy, I_n (ev) the ionization energy of the nth state and $\beta_n = \beta_a$ the atomic luminous efficiency (Table L). For higher ionized states, not covered by the table, an empirical-theoretical expression[26] can be used,

$$\beta_n = 46n^3/I_n^4, \tag{8–9}$$

which involves a logarithmic probable error of ± 0.22.[26]

In Eq. 8–8, the main source of uncertainty is in P_n; this depends upon Π, the probability of ionization + excitation, and is calculated by numerical integrations according to the rules of Sections 7.7, 7.8, and 7.3, the deceleration of the atom being effected essentially through elastic collisions. The results are contained in the upper portion of Table LI (β_1) which, as compared with an early calculation,[7] incorporates the following improvements or alterations: *1*, the use of the sensitivity curve of the human eye according to Table XLIX, instead of a uniform spectral band 4500–5700 A; *2*, the use of the luminous efficiencies β_a (Table L) for the neutral and singly ionized atomic states, instead of a merely statistical relation similar to Eq. 8–9; *3*, the introduction of an "initial excitation," assumed to take place for each meteor atom in collisional breakup of the vapor molecules into atoms (e.g. $MgO \rightarrow Mg + O$).

The first two alterations are trivial. As to the third, in the many-body process of collisional dissociation, excitations are more favored than in two-body collisions. It is assumed here that meteor atoms originating in dissociation are in an excited state. This increases the luminous efficiency at low velocities, little affecting that at high velocities; it removes a discrepancy between the observed and theoretical variation of meteor stream intensity with apical direction[8] for retrograde meteors (which are expected to move at random relative to the sun), and therefore appears to be justified by statistical observations.

In Table LI the second line gives the estimated degree of

TABLE LI

Visual Radiative Efficiency of Meteor Impact Radiation

(same for stone and iron; β = fraction of initial kinetic energy converted into visible light, in units of 0.001; w = initial velocity, km/sec)

w	5.2	7.4	10.4	14.8	20.9	29.6	41.8	59.2	83.6
Degree of dissoc.	0	0	$\frac{1}{8}$	$\frac{1}{2}$	1	1	1	1	1
1. Greatly diluted coma (∂_0 small)									
Initial, β_i	0	0	0.28	0.55	0.55	0.28	0.14	0.07	0.04
Impact, β_c	0	0.17	0.43	0.45	0.41	0.36	0.31	0.30	0.26
Total, $\beta = \beta_1$	**0.00**	**0.17**	**0.71**	**1.00**	**0.96**	**0.64**	**0.45**	**0.37**	**0.30**
2. Compact coma (∂_0 large)									
Initial, β	0	0	0.28	0.55	0.55	0.28	0.14	0.07	0.04
Impact, β	0	0.04	0.25	0.55	0.91	1.39	1.96	2.53	3.07
Total, $\beta = \beta_2$	**0.00**	**0.04**	**0.53**	**1.10**	**1.46**	**1.67**	**2.10**	**2.60**	**3.11**

dissociation of the vapor molecules as depending on velocity; this determines the "initial β" (β_i); "impact β" (β_c), calculated according to Eq. 8–8, is the component of luminous efficiency derived from collisional excitation of the moving atom. β_1, in units of 10^{-3}, is the total efficiency for a diluted coma, i.e., for single atoms decelerating in air without encounters with other meteor atoms.

The second section of the table gives β_2, the luminous efficiency in units of 10^{-3} for a compact cloud of meteor vapors, feebly diluted in the beginning; the cloud expands while incorporating air molecules (atoms) and decelerating *en bloc* at the same time; its degree of dilution also increases, in direct relationship to deceleration (Fig. 14B.). Secondary collisions frequently take place between meteor atoms and lead to appreciable radiation, especially at high velocities.

One half, or more of the radiation is evolved during the first half-energy range of deceleration of the meteor atoms, and the degree of dilution $\partial_0 = \partial_\beta$ which decides the transition between Cases β_1 and β_2 can be assumed to correspond to the half-energy phase. Interpolation between β_1 and β_2 of Table LI can be made according to

$$\beta_0 = (\beta_1 + \beta_2 \partial_\beta)/(1 + \partial_\beta). \qquad (8\text{–}10)$$

For dustballs, ∂_β is to be taken according to the first section (half-energy phase) of Table XLV. For compact meteoroids in normal environment the dilution factors as for half-energy range are calculated from Eqs. 7–17 and 7–18 as given in Table LII.

Tables XLV, LII, and LI with Eq. 8–10 enable us to calculate luminous efficiencies of dustballs and compact meteors in normal environment. Although the absolute efficiencies so calculated may be systematically in error by as much as in a ratio of 1 to 2, the refinement makes it possible to take into account differential effects and is therefore justified. Thus, for a dustball of mass 5 g and velocity 30 km/sec Table XLV yields $\partial_\beta = 0.06 \times 5 = 0.30$; from

Table LI, $\beta_1 = 0.64 \times 10^{-3}$, $\beta_2 = 1.67 \times 10^{-3}$, and β_0 (impact) $= 0.88 \times 10^{-3}$, according to Eq. 8–10.

TABLE LII

Dilution Factor Ratio, $\eth_\beta/r_0^4$, for Radiation Phase of Average Coma of Compact Meteoroids in Normal Environment

E_r, ev	410	205	102.5	51.2	25.6	12.8	6.4
			1. Stone; $r_0^4 = 0.0290M^{4/3}$				
w, km/sec	95.6	67.7	47.8	33.9	23.9	16.9	12.0
$\bar{z}^2$	22.8	14.5	8.5	6.9	6.3	2.37	2.04
$\eth_\beta/r_0^4$	3.8	11.9	40	100	220	1170	2700
			2. Iron; $r_0^4 = 0.00958M^{4/3}$				
w, km/sec	83.6	59.2	41.8	29.6	20.9	14.8	10.4
z^2	7.3	4.8	3.2	3.2	3.2	1.2	1.0
$\eth_\beta/r_0^4$	14.8	48	140	290	580	3100	7400

Similarly, for compact iron of $w = 42$ km/sec, $r_0 = 0.5$ cm, from Table LII $\eth_\beta = 140 \times (0.5)^4 = 8.8$; from Table LI, $\beta_1 = 0.45 \times 10^{-3}$, $\beta_2 = 2.10 \times 10^{-3}$, and β_0 (impact) $= 1.93 \times 10^{-3}$.

8.3 Temperature Radiation in Discrete Energy Levels

In Eq. 8–10 and Table LI only impact radiation produced in the one-sided process of deceleration of primary and secondary atoms is taken into account. However, for large meteors the thermal energy of a compact coma may remain at a high level for a while, and another source of radiation may arise, originating in the thermal collisions of the atoms. Although thermodynamic equilibrium of electrons is not attainable, the kinetic energies of the atoms will approach the condition of equipartition. Radiation in discrete energy levels arises from inelastic collisions between the coma atoms. This "temperature radiation" is produced without the participation of electrons, on account of underexcitation. The total radiation depends on the lifetime of the coma and its atomic temperature, or kinetic energy content T'.

Let us follow up the life history of the kinetic energy $E = \frac{1}{2}w^2$, carried by unit mass of the meteor substance at the moment of vaporization. E' is the thermal energy of the coma per g of meteor substance, T' the thermal energy per unit mass, ψ the fraction of energy lost inelastically, ∂ the dilution factor or the fraction of air to meteor substance, w' the translational velocity of the coma. It can be shown that, from momentum and energy conservation principles, $w' = w/(1+\delta)$, $E' = E(1-\psi)\partial/(1+\partial)$ and $T' = E'/(1+\partial)$ (erg/g of coma). When the inelastic fraction is small, $\psi \to 0$, with increasing dilution T' increases to a maximum T'_m at $\partial = 1$ and then decreases again. For order-of-magnitude calculations, we may assume that the temperature radiation equals that emitted by a coma of constant temperature T'_m during a time interval equal to that of expansion to $\partial = 1$; the overestimate of temperature will be nearly balanced by the underestimate of the lifetime. We note that at $\partial = 1$ and $\psi \to 0$, $E' = \frac{1}{2}E$, $T' = \frac{1}{4}(\frac{1}{2}w^2)$, and the gaskinetic velocity of the coma atoms equals about one-half the velocity of the meteor.

The velocity of expansion of the coma is close to $[2/(3\pi)]^{\frac{1}{2}}v_2$, the mean relative velocity of two atoms in the coma is $4v_2/(3\pi)^{\frac{1}{2}}$, where $v_2 =$ mean square molecular velocity. The ratio of the two is $2\sqrt{2}$. Hence, during the time of expansion of the coma from 0 to a radius Y at $\partial = 1$, each atom in the coma will sweep a zig-zag path equal to $2\sqrt{2}\,Y$. The number of "half-energy collisions" per atom of the coma during its conventional lifetime thus equals

$$\nu_c = \sqrt{2}\,Y(1+\partial)/\lambda, \tag{8–11}$$

where a factor of $\frac{1}{2}$ is included because every collision is counted twice; $\lambda/(1+\partial)$ is the half-energy range in the coma and λ that in the free atmosphere. Also, by definition of $\partial = 1$, the mass or air swept by radius Y over 1 cm of path equals that of the meteor substance vaporized per 1 cm,

$$\pi\rho Y^2 = M/L, \tag{8–12}$$

where M is the mass, L the effective length of the path of ablation of the meteor.

The average velocity of a molecular encounter is $\sqrt{2}$ times the gaskinetic velocity. Let the value of β_c, or the primary impact efficiency (4th line of Table LI) at $\frac{1}{2}\sqrt{2}\,w$, the gaskinetic velocity in the coma, be β'_c; one half-energy collision in the coma will convert a fraction of about $\frac{3}{4}\beta'_c$ of the kinetic energy into visible light, if it takes place between meteor atom and air atom, twice as much if both colliding particles are meteor atoms, and zero if both are air atoms; the average is $\frac{3}{4}\beta'_c$.

In Eq. 7–27, with the postulate of optimum action, E_r/Ω is only slightly greater than 1, and $k_i \simeq 0.03$ is thus the ratio of the number of inelastic to all collisions. Elastic collisions prevail. For equal colliding masses, $m = 1$, Eqs. 7–4 and 4–49 yield $V_0 = \frac{1}{4}E$ for a half-energy elastic collision; this is also the inelastic energy transfer for same target radius. For a molecular encounter, E is the double of the average kinetic energy. Thus, for each half-energy collision (97 % elastic), a fraction $\varepsilon = \frac{1}{2}k_i = 0.015$ of the thermal energy will be lost into inelastic transitions. The total inelastic energy fraction after ν_c collisions with the half-energy target radius is thus

$$\psi = 1 - \exp(-\varepsilon\nu_c). \tag{8–13}$$

For one average half-energy collision per atomic pair in the coma, the total inelastic energy is 0.015 E'; the fraction converted into visible light is $\frac{3}{4}\beta'_c(2E')$. Hence visible light amounts to a fraction of $100\beta'_c$ of the total inelastic energy $\psi E'$; with $E' = \frac{1}{2}E(1 - \psi)$ at $\partial = 1$, the radiated fraction of E, or the luminous efficiency of temperature radiation becomes

$$\beta_t = 50\beta'_c\psi(1 - \psi). \tag{8–14}$$

This expression reaches a maximum of $12.5\beta'_c$ at $\psi = \frac{1}{2}$, $\varepsilon\nu_c = 0.69$, $\nu_c = 46$. The derivation, however, was based on the assumption of ψ being small. From the nature of the problem it is clear that β_t should approach asymptotically an upper limit for $\psi \to 1$, instead of dropping to zero according to Eq. 8–14. We assume that our approximation is valid only until $\psi = \frac{1}{2}$, and that for $\psi > \frac{1}{2}$, $\beta_t = 12.5\beta'_c$.

With $\lambda = 10^{-8}\mathfrak{a}/\rho$, where $\mathfrak{a} = 10^8\rho\lambda$ is taken directly as printed from Table XLIV, and with $\partial = 1$, Eqs. 8–12 and 8–11 yield

$$\nu_c = 2\sqrt{2} \times 10^8(M\rho/\pi L)^{\frac{1}{2}}/\mathfrak{a}, \tag{8–15}$$

where ρ is the atmospheric density corresponding to middle trail. Equations 8–14 (with restrictions imposed), 8–13, and 8–15 with $\varepsilon = 0.015$ solve the problem.

For compact iron (Case 4) or stone (Case 7 of Table XXII) in normal environment, after the usual substitutions, a formula whose numerical coefficient is by chance the same in both cases is obtained:

$$\varepsilon\nu_c = 3.8 \times 10^6 r_0{}^2/(\mathfrak{a}w) \text{ (compact).} \tag{8–16}$$

For dustballs, with $\rho = 2 \times 10^4/w^2$,

$$\varepsilon\nu_c = 2.35 \times 10^5 M^{\frac{1}{2}}/(\mathfrak{a}w) \text{ (dustballs).} \tag{8–17}$$

TABLE LIII

Inelastic Collision Frequency and Temperature Radiation from Coma in Normal Environment

(when $\varepsilon\nu_c < 0.69$, use Eqs. 8–13 and 8–14; when $\varepsilon\nu_c > 0.69$, set $\beta_t = 12.5\beta'_c$)

w, km/sec	7.4	10.4	14.8	20.9	29.6	41.8	59.2	83.6
β'_c, 10^{-3} units	0	0.17	0.43	0.45	0.41	0.36	0.31	0.30
1. Compact iron								
$\mathfrak{a}$	3.8	4.5	4.7	7.2	7.6	8.2	9.3	10.9
$\varepsilon\nu_c/r_0{}^2$	1.34	0.88	0.54	0.25	0.17	0.109	0.069	0.042
2. Compact stone								
$\mathfrak{a}$	3.6	4.4	4.5	6.6	7.7	10.7	14.6	17.2
$\varepsilon\nu_c/r_0{}^2$	1.43	0.84	0.57	0.27	0.16	0.084	0.044	0.026
3. Dustballs								
$\mathfrak{a}$	3.6	4.4	4.5	6.6	7.7	10.7	14.6	17.2
$\varepsilon\nu_c/M^{\frac{1}{2}}$	0.088	0.052	0.035	0.017	0.0105	0.0052	0.0027	0.0016

Table LIII contains auxiliary data for the calculation of β_t. Thus, for a dustball of $M = 1$ g, $w = 41.8$ km/sec, $\varepsilon v_c = 0.0052 \cong \psi$, or $\beta_t = 0.36 \times 10^{-3} \times 5.2 \times 10^{-3} = 2 \times 10^{-6}$, a negligible quantity. For a compact stone of equal mass and velocity, $r_0 = 0.4$ cm, $\varepsilon v_c = 0.014$, $\beta_t = 5 \times 10^{-6}$. Temperature radiation is prominent only in massive meteors, and is more important for compact meteoroids than for dustballs of equal mass.

8.4 Other Sources of Meteor Radiation

The luminous efficiencies derived in Sections 8.2 and 8.3 are based on the assumption that all the substance of the meteor is ultimately vaporized. The smallness of the residual mass (Section 5.5) justifies this assumption with respect to meteors of the visual and telescopic range. However, in micrometeors a considerable fraction of the meteoroid survives, yet it would be incorrect to assume that the luminous efficiency will be smaller in proportion to the fraction of mass actually vaporized. The air molecules impinging on the surface of the meteoroid will lead to "cold" impact radiation, a kind of fluorescence including discrete energy levels and continuous radiation. The case of the Pultusk meteorite (Section 2.3) shows the importance of surface impact radiation. Its efficiency per collision is difficult to estimate, but it appears probable that it is not less than that in collisions of free atoms. The material of a meteoroid which does not evaporate will have to be decelerated by the same number of collisions with air molecules as one that evaporates and loses its momentum in a diffuse state; therefore, the luminous efficiency of the nonevaporating part of the meteor mass will not be less than that of the evaporated portion.

Not being able at present to estimate the luminous efficiency at impact on solid (or liquid) surface, the best thing to do is to assume for the luminous efficiencies of micrometeors the values of β_1 as given in Table LI.

Another source is thermal radiation from the heated surface of the meteoroid. Wien's approximation of Planck's radiation formula, applied to visual light of an average wavelength 0.505 μ and practically valid for $T < 10000°$ K, gives the fraction of energy comprised in a limited spectral region as $b_v = T^{-4} \exp(-28500/T) \times \text{const}$, where the constant depends upon the spectral width of the region.

For solar radiation, $T = 5800°$ K, $b_v = 0.117$ is found from the integration of the sensitivity curve of the human eye over the solar energy curve;[26] this determines the constant, whence the luminous efficiency of black-body radiation at temperature T is found as

$$b_v = 1.78 \times 10^{16} T^{-4} \exp(-28400/T). \qquad (8\text{–}18)$$

For $T = 2100°$ K, $b_v = 1.25 \times 10^{-3}$.

In notations of Section 6.8, and according to Eqs. 4–27 and 4–25, the total kinetic energy dissipated per second is $\frac{1}{2}\xi w^2 + 2Q/\gamma$; the energy converted into visual black-body radiation is $b_v Q$. Hence the contribution of black-body radiation to luminous efficiency becomes

$$\beta_b = b_v \Theta / (\tfrac{1}{2} w^2/h + 2\Theta/\gamma). \qquad (8\text{–}19)$$

Table XXXII gives $\Theta = 0.19$ at $T = 2100°$, and for $w = 2 \times 10^6$ cm/sec, $h = 8 \times 10^{10}$ erg/g, we find $\beta_b = 10^{-5}$ or a negligible value. Although b_v increases with temperature, Θ decreases, and the resulting contribution, Eq. 8–19 remains small.

With the correction for black-body radiation, the total luminous efficiency becomes

$$\beta = \beta_0 + \beta_t + \beta_b, \qquad (8\text{–}20)$$

where β_0 is given by Eq. 8–10, β_t by Eq. 8–14, and β_b by Eq. 8–19. For most visual meteors β_0 is the only important component.

8.5 Calculation of Meteor Masses

The luminous efficiency, β, together with the observed magnitude, $\mathfrak{M}$ (cf. Appendix 3), are at present the most reliable means of determining the masses of meteors. Estimates of magnitude of a meteor made by different observers involve a probable error of about $\pm$ 0.5 mag between 1st and 4th magnitude; for bright meteors (negative magnitudes, brighter than Sirius) the error increases to $\pm$ 1.0 mag. Photographic determinations of meteor magnitudes may be more accurate, although large systematic errors due to "reciprocity failure" may be involved. For statistical purposes, only visual magnitudes are at present available.

With the sensitivity curve of the human eye (b as in Table XLIX), the photometrically determined standards of stellar magnitude referred to the sun, and the visual radiation of the sun in "visual ergs" per cm² and sec as defined by

$$j_\odot = \int b_l P_l \, dl,$$

the intensity of radiation of a meteor, j, in visual ergs per second,[8] is given by

$$\log j = 9.72 - 0.4\mathfrak{M}_0. \tag{8–21}$$

Here $\mathfrak{M}_0$ is the absolute magnitude of the meteor, i.e., its stellar magnitude at a distance of 100 km in the direction of the zenith; $P_l \, dl$ is the radiation of the sun in the zenith in erg/cm² · sec between wavelengths l and $l + dl$; the constant 9.72 in Eq. 8–21 corresponds to the sun's stellar magnitude being $-$ 26.72 and to the energy distribution of solar radiation according to the results of the Astrophysical Observatory of the Smithsonian Institution.

The observed magnitude of a meteor is a certain average over its visible trajectory; thus, j is also an average, to be considered as covering the time of visibility, t, of the meteor. Hence follows the relation between mass, M(g), and luminosity (j) of the meteor:

$$jt = \tfrac{1}{2}\beta M w^2, \tag{8–22}$$

where β is the luminous efficiency according to Eq. 8–20, and w the velocity in cm/sec. The duration of visibility, or of flight, is given by

$$t = L/w, \tag{8-23}$$

where L is the visible length of the trajectory. Equations 8–22 and 8–23 apply to "ordinary" meteors, not too bright, which show little detail along their trajectory, especially little variation of brightness.

For very bright objects which can be observed at early and late stages of their flight, when they are relatively faint, j is likely to be overestimated; as a compensation, instead of the apparent length L, a certain effective brighter and shorter portion L_1 of the trail should be used in Eq. 8–23, $L_1 < L$.

For photographic observations the product on the left-hand side of Eq. 8–22 can be determined directly from photometric measurements, $jt = \int j\, dL/w$.

For statistical use Eq. 8–22 can be simplified, considering that the absolute length of the trail for dustball-type objects (which appear to be in the majority among visual meteors) should not depend upon mass (if the dustgrain radii are constant) but only on velocity. The Arizona data give[8] $t = 0.67(2.1 \times 10^6/w)^{0.7}$ (sec). Further, with $\beta = \beta_1$ (Table LI) for the dustballs, which can be represented by

$$\beta_1 = 2000/w \text{ (for } w = 1.5\text{–}8.0 \times 10^6 \text{ cm/sec)}, \tag{8-24}$$

Eq. 8–22 after proper substitution becomes

$$\log M = 10.97 - 1.7 \log w - 0.4\mathfrak{M}_0 \text{ (dustballs)}, \tag{8-25}$$

where w is in cm/sec and M in grams. Its validity is limited to about $M < 1$ g. When these approximations are not acceptable, the general formula with β as from Eq. 8–20 should be used:

$$\log M = 10.02 + \log L - \log \beta - 3 \log w - 0.4\mathfrak{M}_0. \tag{8-26}$$

CHAPTER 9

Some Applications

9.1 Draconids as Dustballs

With the display of the Draconids in October, 1946, it was realized for the first time that the concept of compact meteoroids may fail completely.[45] The air mass traversed by these meteors was so small that, for complete evaporation, the radii of their compact meteoroids must have been smaller than 0.01 cm, whereas the meteor masses were of the order of 1–100 grams as calculated from the luminous efficiency. The discrepancy can be removed by assuming that the meteors were aggregates of thousands or millions of dust-grains, loosely bound and dissolving into grains when entering the earth's atmosphere.

In the following, upper limits to the grain radii of the Draconids are calculated on the assumption that the visible trail of the meteor is identical with the ablation path of each of its individual grains. This will be true if the dustball disintegrates before the observed beginning of its trail. If the disintegration continues along the trail, the paths of the grains will be shorter, and their radii smaller than calculated.

Table LIV contains the results of calculation for the photographic observational material obtained by Jacchia, Kopal, and Millman.[45] The first five lines contain the average observational data: $\mathfrak{M}_0$, the absolute visual magnitude (i.e., reduced to a distance of 100 km) averaged over the trail (cf. Section 8.5); n, the number of meteors; H_a, H_d, the altitudes of appearance and disappearance; t, the duration of observed flight.

In the sixth line, r_0' is the upper limit of grain radius,

TABLE LIV

Draconids of 1946 as Dustballs
(dustgrain radii, luminous efficiencies, and total masses)

$\mathfrak{M}_0$	0.3	−0.5	−1.4	−2.4	−3.5	−4.5	−5.6
n	13	44	42	45	15	8	4
H_a, km	95.1	96.9	98.4	97.7	98.8	98.6	98.4
H_d, km	90.2	91.0	91.0	89.5	87.6	85.3	86.5
t, sec	0.32	0.39	0.49	0.54	0.74	0.80	0.78
r_0', 10^{-3} cm,	$\leqq$ 5.4	4.6	4.7	5.9	8.9	13.5	10.6
X_e	0.66	0.78	0.86	0.92	0.95	0.97	0.98
$\mathfrak{d}_0$	0.010	0.023	0.058	0.13	0.37	0.79	1.7
β_0, 10^{-3}	0.87	0.87	0.90	0.93	1.04	1.15	1.26
β_t, 10^{-3}	0.22	0.34	0.54	0.79	1.27	1.79	2.51
β, 10^{-3}	1.09	1.21	1.44	1.72	2.31	2.94	3.77
M, g	0.46	1.06	2.6	5.9	17	36	74
$N' \geqq$	2×10^5	8×10^5	2×10^6	2×10^6	2×10^6	10^6	4×10^6

calculated according to the rules of Section 5.8. The radius is so small that, according to the chart of ablation for stone (Table XXXIX), Case 2 of isothermal melting and subsequent vaporization is the only one to be considered, nor is there any possibility of flattening or breakup of the drops. However, the trails of the Draconids were rather short, suggesting abnormal environment. Therefore, eliminating the prehistory of melting which determines the first term in the brackets of Eq. 5–37, we take the difference of the two expressions 5–36 for the beginning and end of the visible trail; the beginning, identified as the start of vaporization, requires $X = 0$, $\rho = \rho_a$; for the end, with $\rho = \rho_d$, we set $X = X_e$. Hence

$$\rho_d - \rho_a = 2hX_e r_0' \delta \cos Z/(J\gamma a w^2). \qquad (9\text{–}1)$$

This differential expression is valid for arbitrary environment in a homotropic atmosphere, although the equations from which it has been derived only apply to normal environment.

The parameter of ablation at the apparent end point, X_e, as given in the 7th line of the table, is estimated from the limiting magnitude of the photographic plate according to the method of Section 5.4. The velocity of the Draconids is 23.1 km/sec; however, dustgrains are expected to decelerate and, according to Section 5.5, an average velocity of $w = 2.21 \times 10^6$ cm/sec for the visible path is assumed. As to γ, according to Eq. 5–21, radiation losses would influence it considerably; in notations of Section 6.8, we set

$$\gamma = 0.8/(1 + \Theta). \tag{9–2}$$

For the mean height and velocity of the Draconids Tables XXXIII and XXXII suggest $T = 1947°$, $\Theta = 0.43$, whence $\gamma = 0.560$. With ρ_a and ρ_d as from the Rocket Panel (Table I), $J = 0.25$ as for a spherical drop, $h = 6.05 \times 10^{10}$ erg/g as for pure evaporation, $a = 6.8 \times 10^5$ cm, $\delta = 3.4$ and $\cos Z = 0.661$ as an average, from Eq. 9–1 the values of r_0' of Table LIV are calculated. These are upper limits of grain size; the actual radii may be considerably smaller for two reasons. First, the Rocket Panel densities are likely to be overestimated[20] and, thus, also the values of r_0' as defined by Eq. 9–1. Second, although theory indicates a noticeable deceleration of the dustgrains, almost no measurable deceleration could be detected in the Draconids;[45] this can be understood only by assuming that their visible trails were not identical with the paths of ablation of the individual grains; apparently, the dustballs were shedding off grains all along the visible trail without being appreciably decelerated. In such a case, the trajectories of the grains must have been very much shorter than those of the parent dustballs, and their radii smaller than calculated. An upper limit of $r_0' < 0.002$ cm is thus suggested.

The 8th to 13th lines of Table LIV contain the results of calculation of the total masses, according to the rules of Chapter VIII. The luminous efficiency depends upon mass, and the mass upon luminous efficiency; the solution is therefore obtained by trial and error until consistent values of β and M are obtained. The luminous efficiency as defined by Eq. 8–20 essentially consists of two components only, β_b being negligible. In the Table, ∂_0 is the dilution factor according to

Table XLV; β_0 the luminous efficiency of impact radiation according to Eq. 8–10 and Table LI; β_t that of temperature radiation according to Eqs. 8–14, 8–13, and Table LIII; β the total luminous efficiency, all three in units of 0.001; M the total mass in grams, calculated according to Eq. 8–26; N' the minimum number of grains in a dustball. In the calculation of β and M the initial velocity, $w = 2.31 \times 10^6$ cm/sec, has been used and not its decelerated value.

The data of Table LIV refer to bright and massive meteors of the photographic range; it is over this range that marked changes in the dilution factor and temperature radiation are taking place.

9.2 The Average Naked-Eye Meteors

The Arizona material represented in Table XVIII is typical. As has been already pointed out, the altitude range, 0.65–0.75 of that in homotypic ablation (Cases 4, 5, 6, or 7), excludes homotypic ablation and leaves only Case 2 for stone as the only possible one; the altitude range for Case 2, equal to 0.75 of that in homotypic ablation (cf. l in Table XXII), is almost identical with that for the Arizona meteors. Thus, as a rule, the ablating meteoroids must undergo complete fusion, previous to their vaporization; this conclusion, based on a model of normal environment, becomes even more stringent if the meteoroids are dustgrains in abnormal environment, of small size and ablating over a path which is but a fraction of the visible trail.

As in the case of the Draconids, the upper limit of the grain radius, r_0', can be calculated on the assumption that the path of ablation of each grain is identical with the visible trail. In Table LV the values of r_0' so defined are calculated from Eq. 5–36 with $\rho = \rho_d$ as for the point of disappearance according to the Rocket Panel (Table I), ϕ as for Case 2 in Table XXI, and $X = X_e$ for the point of disappearance as given in Table XVII.

On the other hand, the absence of foreshortening of the trail as compared with Case 2 indicates that the molten

TABLE LV
Arizona Visual Meteors

Group I, $w \cong 68$ km/sec (retrograde orbits)						
$\mathfrak{M}_0$, mag	3.38	3.05	2.46	1.77	0.76	−0.79
M, g	0.010	0.014	0.024	0.044	0.11	0.47
r_0' cm, $\leqq$	(0.071)	0.043	(0.065)	(0.076)	0.034	(0.224)
$N' \geqq$	5(2)	12	15(6)	20(7)	190	79(3)
Group II, $w \cong 42$ km/sec (direct orbits)						
$\mathfrak{M}_0$, mag	3.52	3.19	2.52	1.94	1.11	−0.39
M, g	0.020	0.027	0.051	0.086	0.18	0.74
r_0' cm, $\leqq$	(0.067)	0.042	(0.089)	(0.104)	(0.125)	(0.502)
$N' \geqq$	10(5)	27	26(5)	43(5)	92(7)	350(0.4)
Group III, $w \cong 21$ km/sec (direct orbits of small eccentricity and inclination)						
$\mathfrak{M}_0$, mag	3.64	3.35	2.76	2.05	1.31	−0.37
M, g	0.058	0.076	0.13	0.25	0.50	2.35
r_0' cm, $\leqq$	0.032	0.025	0.027	0.038	(0.084)	(0.159)
$N' \geqq$	130	340	480	330	250(60)	1100(41)

drops are not flattened, whence, according to Table XXVII, the surface tension criterion sets an overriding condition for the upper limit of radius,

$$r_0' \leqq 0.052 \text{ cm}. \tag{9–3}$$

This limit does not depend on the assumed absolute density of the atmosphere. In Table LV, the calculated values of r_0' which exceed this limit are bracketed; in these cases $r_0' \leqq 0.052$ cm is to be assumed.

The other data given in Table LV are: $\mathfrak{M}_0$, the absolute magnitude; M, the total mass in grams, calculated from Eq. 8–25, i.e., as for a greatly diluted coma and negligible temperature radiation; and N', the minimum number of grains in a dustball. The bracketed values of N' are those corresponding to the bracketed cases of r_0'; the figure given without brackets then corresponds to $r_0' = 0.052$ cm. As can be seen from the example of the Draconids, temperature radiation is indeed expected to be small except for the group of brightest magnitude (last column in Table LV) where it may amount to some 20 % of the total; the use of Eq. 8–25 is thus justified. Table LV is arranged into the same groups as Table XVIII from which the other relevant data can be inferred.

In all cases but one even the bracketed values of N' exceed unity; these being absolute lower limits, the conclusion is that, as a rule, in their majority visual meteors are aggregates of a number of grains. On the other hand, the values of N' in Table LV are very much smaller than for the Draconids; although no definite conclusion can be drawn—the actual grain size may be very much smaller than the calculated upper limit—the elongation of the luminous image of a meteor toward the end point[46] would indicate that many grains travel, indeed, over a considerable fraction of the visible path; in such a case, the calculated values of r_0' will be within the order of magnitude of the true values, and the grains of average sporadic dustball meteors, of the order of $r_0' = 0.01$–0.03 cm, appear to be indeed coarser than those of the Draconids.

The grain radii as quoted in Table LV are based on Rocket Panel atmospheric densities, on a high value of $\gamma = 0.8$, and on undecelerated velocities. The average velocities are themselves not accurate, and may be lower by as much as 10 per cent.[8, 9] Assuming one-half the Rocket Panel densities, $\gamma = 0.6$ as reduced by radiation losses (Eq. 9–2) and a velocity by 5 % smaller, the calculated radii will be made 3 times smaller and N' (bracketed) increased 27 times. Thus, even the only contradictory case (Group II, last column) will be changed from $N' > 0.4$ to $N' > 10$, and the contradiction will be removed.

There are other convincing statistical reasons[8, 9] for accepting the dustball nature of the majority of visual meteors; there is hardly any doubt left in this respect.

The knowledge of the true size of the dustgrains may have important cosmogonic implications. The photographic study of the spreading of the cluster of grains, as the result of unequal deceleration of drops of different radii, may give a clue to the actual grain radii; such study can be performed on trails interrupted by a rotating shutter, or with the aid of vibrating cameras.[46] For a bright Leonid meteor, of total mass of 2.4 g, the author[30] found in such a manner $r_0' =$ 0.02–0.03 cm as the most frequent grain size and 0.031 cm as an upper limit, with the Rocket Panel atmospheric densities as a basis.

APPENDIX 1

Physical Characteristics of Iron

Density 7.8. Molecular and atomic weight 56. *Atomic mass* 9.30×10^{-23} g. *Atomic volume* 1.192×10^{-23} cm^3. Cube root of atomic volume 2.284×10^{-8} cm.

Nickel-iron: body-centered cubic lattice (α-iron, kamacite), linear parameter of lattice (two interposed cubic systems) $b = 2.876 \times 10^{-8}$ cm, minimum distance between atoms (eight bonds) $\frac{1}{2}\sqrt{3}\,b = 2.490 \times 10^{-8}$ cm; face-centered cubic lattice (γ-iron, taenite), linear parameter of lattice (four interposed cubic systems) $b = 3.622 \times 10^{-8}$ cm, minimum distance between atoms (twelve bonds) $b/\sqrt{2} = 2.562 \times 10^{-8}$ cm.

Lattice energy: 96600 cal/mole or 4.18 ev per atom [= total heat of vaporization from 0° K to T_b (temp. of vaporization) *minus* heat content of vapor at T_b $(c_p T_b)$]; 1.04 ev per bond and atom for α-iron, and 0.70 for γ-iron.

Melting point 1800° K (with 10 % Ni, 1770° K), *boiling point* 3508° K ($\pm$).

Average specific heat: solid 6.91×10^6 erg/g · deg (0°–1500° C); liquid $= 6.66 \times 10^6$ erg/g · deg; gaseous $c_v = 2.23 \times 10^6$ erg/g · deg, $c_p = 3.72 \times 10^6$. *Latent heat of fusion* $= 2.69 \times 10^9$ erg/g. *Latent heat of vaporization* 6.40×10^{10} erg/g = 3.72 ev/atom.

Vapor Pressure, d/cm^2: iron, $\log p_v = 10.607 – 16120/T$; nickel, $\log p_v = 10.725 - 16120/T$.

Evaporation from unshielded surface: $\xi = 3.26 \times 10^{-4}\, pT^{-\frac{1}{2}}$ (g/cm^2 · sec; p in d/cm^2; T° K); interpolation formulae $\log \xi = 5.167 - 15700/T$ ($T > 2200$° K); $\log \xi = 6.167 - 17900/T$ ($T < 2200$° K).

Surface tension: 1200 d/cm[60].

Coefficient of viscosity 0.026 to 0.019 poises (d · sec/cm^2) for iron melt with 1 % C between $T = 1673$ and 1773° K; 0.023 to 0.016 for 2.5 % C for the same limits of temperature of melting. $\eta = 0.02$ can be assumed for molten meteoric iron, decreasing to 0.01 poise at the effective temperature of vaporization.

Thermal conductivity: probable value $k_t = 4 \times 10^6$ erg/cm^2 · sec · (deg/cm) for meteoric iron (lower than for pure iron).

Nickel content of meteoritic iron: average 8.5 % (5–50%). In octahedrites (6–13 % Ni; most common variety of meteorites): kamacite, aver. 6 % Ni; taenite, 30 % Ni; mixture of these in different proportions determines the total nickel

Strength of Meteoritic and Other Iron[61]

	Brinell hardness	Tensile strength 10^{10} d/cm^2	Compressive strength 10^{10} d/cm^2
Canyon Diablo kamacite	230	0.83	1.5
taenite	510	1.8	3.4
Henbury kamacite	170	0.61	1.1
taenite	210	0.76	1.4
Troilite (FeS)	180	0.65	1.2
Hexabedrites	160	0.58	1.1
Hard meteoritic inclusions (schreibersite Ni, Fe, P; cohenite—Ni, Co, Fe, C)	950	3	6.3

Tentative Correlation with Nickel Content[61]

Total nickel %	0	5	7	10	20	30	45
	Brinell hardness						
Meteorites, kamacite (α-Fe)	90	150	170	200	300	380	—
Meteorites, taenite (γ-Fe)	—	—	90	130	235	380	660
Artificial alloys (α-Fe)	140	200	220	245	340	—	—

content. For more than 27 % Ni all is taenite. Widmanstätten crystallic pattern appears in octahedrites; it consists of taenite faces cemented by kamacite filling.

Hexahedrites ($<6\%$ Ni; rare) have no distinct crystal structure, same as ataxites ($>13\%$ Ni, rare).

Compressibility: bulk modulus (all-sided compression) 1.67×10^{12} d/cm^2; Young's modulus (tensile) 2.0×10^{12} d/cm^2.

The reflectivity of metallic meteorites as determined in the laboratory[62] is 58 % (total albedo). In space the reflectivity may be lower by an order of magnitude, owing to pitting by micrometeors, corpuscular radiation, and cosmic rays.

Except for nickel, cobalt, chromium, and sulfur, the other elements occur in iron meteorites only in traces; sulfur appears only in troilite inclusions (FeS).

Physical Characteristics of Meteoritic and Terrestrial Stone

Meteoritic stone: density variable, average $\rho = 3.4$; mainly ferromagnesian silicates, olivines with different proportion of Fe and Mg; Ca, Al silicates in smaller proportions.

Typical minerals		Density	Melting or decomp. point, °K	
Forsterite	Mg_2SiO_4	3.22	2178 melt.	Artificial olivine
Fayalite	Fe_2SiO_4	4.28	1478 melt.	,, ,,
Fayalite	,, +impur.	—	1481 melt.	Natural olivine
Hematite	Fe_2O_3	5.25	1838 melt.	—
Ferrous oxide	FeO	5.7	1693 melt.	—
Magnetite	Fe_3O_4	5.17	1811 decomp.	—
Magnesia	MgO	3.60	3070 melt.	—
Christobalite	SiO_2	2.32	1990 melt.	Artificial
Quartz	SiO_2	2.65	—	Natural
Troilite	FeS	4.84	1470 melt.	—
Pyroxene			1248 decomp.	Artif. and natur.
(Acmite)	$Na_2O . Fe_2O_3 . 4SiO_2$	3.56	(Fe_2O_3)	
,,	,,	—	1548 melt.	,, ,, ,,
Lime	CaO	3.42	2840 melt.	,, ,, ,,

The average melting point of meteoritic stone may be around 1800° K; the precise value is of little importance, the material being extremely viscous at the temperature of fusion and becomes actually fluid only at higher temperatures.

Composition of Olivine Samples (Terrestrial), %:

	SiO_2	Al_2O_3	Fe_2O_3	FeO	MgO	CaO
Olivine I	49.29	18.49	2.38	6.77	6.09	8.14
Olivine II	49.24	16.81	6.16	3.60	8.02	9.49

Viscosity (extrapolated values are in brackets):

T, °K	1383	1423	1473	1573	1673	2000	2500	3000
				η, poises				
Olivine I	—	37900	3180	296	137	(18)	(2.4)	(1.0)
Olivine II	—	—	732	173	120			
Basalt	3550	—	—	126	55	—	—	—
Techenite (less rich in MgO than basalt)	21500	—	—	1820	851	—	—	—

Average Elementary Composition of Meteoritic Stone (including 5.5 % troilite but without metal inclusions)

Element	O	Mg	Si	Fe	S	Al	Ca	Na	K	Others
% weight	39.8	15.1	20.3	15.4	3.8	1.5	1.8	0.7	0.06	1.5
% atoms	56.6	14.3	16.5	6.3	2.7	1.3	1.0	0.7	0.05	0.6

Mean atomic weight, 23; mean molecular weight of vapors, 50.

Thermochemistry of Molecules:

a. Diatomic molecules

Molecule	O_2	N_2	FeO	SiO	MgO	SO	S_2
Energy of Dissociation, ev	5.08	9.76	4.24	7.2	5.2	4.0 (5.2?)	4.4

Molecule	CaO	AlO	SiN	SiS	MgS	Na_2	C_2
Energy of Dissociation, ev	5.0	3.7	4.5	6.6	2.9	0.73	3.6

b. Other compounds (the data refer to free molecules, and are calculated with a correction for the lattice energy)

Reaction	$Al_2O_3 \rightarrow 2Al + O_2 + O$	$Al_2O_3 \rightarrow 2AlO + O$	$2FeO \rightarrow 2Fe + O_2$
Dissoc. en., ev	(20.0)	(17.7)	3.4
Reaction	$2MgO \rightarrow 2Mg + O_2$	$2CaO \rightarrow 2Ca + O_2$	$SiO_2 \rightarrow Si + O_2$
Dissoc. en., ev	5.3	4.9	9.1
Reaction	$SiO_2 \rightarrow SiO + O$	$Fe_2O_3 \rightarrow 2FeO + O$	$Fe_2O_3 \rightarrow FeO + Fe + O_2$
Dissoc. en., ev	7.0	(5.4)	(4.5)

c. Estimated lattice energies (from Trouton's law, except Fe & Ni)

Molecule	Fe	Ni	Si	SiO_2	MgO	CaO	Al_2O_3	S_2
Lattice en., ev	4.18	4.50	3.6	2.8	4	3.8	4.0	0.8

Composition of Vapors from Meteoric Stone (without metal inclusions):

Molecule	SiO_2	MgO	CaO	Fe	S	O_2	Al_2O_3	Rest	All
% by weight	43.5	25.1	2.5	15.4	3.8	3.8	3.2	2.7	100.0

Vapor Pressures and Boiling Points

	SiO_2 (molten)						CaO			
	Ruff and Schmidt, 1921				Ruff and Kontschak, 1926					
Vapor pressure, mm Hg	23	65	180	760	11	760	10^{-7}	3×10^{-6}	1	B.P.
T, °K	2173	2273	2373	2503	2333	2863	1600	1750	2500	3120

	MgO	Al_2O_3	Fe	S_2	FeO	Fe_2O_3
Vapor pressure, mm Hg	B.P.	53	B.P.	B.P.	(760) Dissociation	(760) Dissociation
T, °K	3070	2850	3508	718	(3500)	(3750)

Weighted mean boiling (or dissociation) *point* 2960° K (without S).

Mean lattice energy per vapor molecule 3.5 ev.

Tentative *vapor pressure* formula: $\log p = 10.600 - 13500/T$ (d/cm²).

Evaporation: $\xi = 3.08 \times 10^{-4} p T^{-\frac{1}{2}}$ (g/cm² · sec; p in d/cm²).

Specific heat (fayalite): Average solid, 0–1200° C, 8.95×10^6 erg/g · deg; liquid, probable 1.1×10^7 erg/g · deg.

Latent heat of fusion, fayalite 2.65×10^9 erg/g. *Latent heat of vaporization*, average for meteoric stone 6.05×10^{10} erg/g (from mean lattice energy).

Surface tension 400 d/cm at fusion, and 360 at temperatures of volatilization ($\pm$).

Thermal Conductivity

Terrestrial rocks	Basalt	Granite	Limestone	Sandstone (dry)
k_t, cal/cm^2 · sec · (deg/cm)	0.005	0.005	0.0025–0.005	0.003

For meteoric stone, $k_t = 2 \times 10^5$ erg/cm · sec · deg (±).

Electric Resistivity of stony meteorites is from 10^4 to 10^8 ohm · cm[63], mostly around 10^5, and similar to that of terrestrial igneous rocks.

Structure: the majority of stony meteorites are chondrites, consisting of cemented granules or chondrules; the achondrites are a less common variety.

Reflectivity of stony meteoritic surfaces is from 0.27 to 0.62[62] (total albedo); pitted rough surfaces in space should have much lower values.

Strength of Some Terrestrial Rocks

	Basalt	Granite	Sandstone	Limestone
Density	3.19–2.94	2.65–2.70	2.2–2.6	2.3–2.7
Porosity (% of pore volume)	0.4–0.5	0.5–1.5	5–20	3–15
Compressive strength, 10^9d/cm^2	2.0–3.5	1.0–2.8	0.5–1.5	0.4–1.4
Shearing strength, 10^8d/cm^2	—	1.5–3.0	0.5–1.5	1.0–2.0
Transverse (flexure) strength, 10^8d/cm^2	—	1.0–2.0	0.25–1.25	0.75–1.25
Tensile strength, 10^7d/cm^2	—	3.0–5.0	1.0–3.0	3.0–6.0

Elastic Characteristics[40]

	Olivine diabase*		Peridotite dunite*		Granite		Quartzitic sandstone	
Pressure, km/cm^2	1	10^4	1	10^4	1	10^4	1	10^4
Bulk modulus, 10^{11} d/cm^2	7.4	8.3	10.6	12.6	4.4	5.5	3.7	4.4
Young's modulus (compr.), 10^{11} d/cm^2	10.1	10.9	14.7	17.4	4.6	7.3	6.5	6.0
Density, g/cm^3	3.00	3.08	3.28	3.29	2.62	2.67	2.64	2.70

* Somewhat similar to meteoritic stone; peridotite is the material of the earth's mantle below 60 km depth.

Miscellaneous Information

Magnitudes:

Stellar magnitude = logarithmic measure of brightness. Increasing magnitude = decreasing intensity of light. Visual magnitudes: Sun = − 26.7; full Moon = − 12, first or last quarter = − 9; Venus = − 4; Jupiter = − 2.5; Sirius = −1.5; Vega, Capella, Arcturus = 0; Pole Star and Stars of the Plough = + 2; faintest stars visible to the naked eye = + 6 mag. Faintest meteors visible = + 5 mag. Visual meteors mostly between 0 and + 4 mag; telescopic meteors from + 5 to + 10 mag; photographic meteors from 0 to − 7; fireballs are brighter than − 1; meteorites from − 10 to − 20 and brighter; micrometeors from + 15 to +25.

The difference of one magnitude corresponds to an intensity ratio of 2.512 (Pogson's ratio); its logarithm is exactly 0.4 by definition. If I_1 and I_2 are the intensities, $\mathfrak{M}_1$ and $\mathfrak{M}_2$ the magnitudes, then

$$\log (I_1/I_2) = -0.4(\mathfrak{M}_1 - \mathfrak{M}_2).$$

Visual magnitudes or intensities are those corresponding to the spectral sensitivity curve of the human eye; there is some ambiguity in definition due to the Purkinje effect, or the variation of spectral sensitivity with intensity.

Photographic magnitude corresponds to the spectral sensitivity of the ordinary (nonsensitized, blue-sensitive) photographic emulsion.

Brinell Hardness and Strength of Materials:

Brinell Hardness (B.H.) = ratio of pressure to area of spherical indentation, kg/mm^2; compressive strength =

about $\frac{2}{3}$ of B.H. For various steels and malleable iron, tensile strength = 0.360 of B.H. (20 specimens, range 0.31–0.46); for cast iron (brittle), tensile str. = 0.104 of B. H. (3 specimens, 0.089; 0.092; 0.134).

Kinetic Energy of Atoms

Velocity, km/sec	KE, ev H	N	Fe
0.92	0.0046	0.06	0.25
1.30	0.0092	0.12	0.50
1.85	0.018	0.25	1.0
2.61	0.037	0.50	2
3.70	0.074	1.0	4
5.22	0.148	2	8
7.40	0.295	4	16
10.44	0.59	8	32
14.8	1.15	16	64
20.9	2.3	32	128
29.6	4.6	64	256
41.8	9.2	128	512
59.2	18.4	256	1024
83.6	36.9	512	2048
118.4	73.8	1024	4096
167.2	147.5	2048	8192

Static Atmosphere:

Pressure gradient versus gravity: $dp = -g\rho\, dH$; $\rho = N\mu$; $p = NkT$; $\rho = \mu p/(kT)$; $dp = -g\mu p dH/(kT)$.

Scale height: $a = kT/(g\mu)$; $dp/p = -dH/a$.

In homogeneous isothermal atmosphere of constant gravity: $a = \text{const}$; $\ln(\rho/\rho_0) = \ln(p/p_0) = -(H - H_0)/a$. Air mass ($A$) per cm² of path ($dx$): $dA = \rho dx$; in oblique incidence (angle Z) $dx = -dH \sec Z$; $dA = -\rho \sec Z \cdot dH = dp \sec Z/g$, valid for a nonhomogeneous, nonisothermal atmosphere.

Here: p = atmospheric pressure; ρ = atmospheric density; H = altitude; g = acceleration of gravity; N = number density of molecules; μ = average molecular mass; k = Boltzmann's constant; T = absolute temperature; Z = zenithal angle of incidence.

Principal Notations

A	Air mass per cm² cross section (A_0)
A'	Quantum-mechanical damping constant
A_s	Air stream intensity
a	Scale height
B	Shape parameter
C	Contour parameter
c_p, c_v	Specific heats
D	Coefficient of diffusion (D_g, D_t)
D_i	Linear dimension (D_1)
d	Kinetic depth (d_v, d_c, d_0, d)
E	Kinetic energy ($E_1, \Delta E, \Delta E_1, \Delta E_i$)
E_r	Relative kinetic energy ($\bar{E}_r$)
f	Aerodynamic drag force (f_a)
g	Acceleration of gravity ($\bar{g}$)
H	Altitude or height (H_a, H_d, $\bar{H}$, H_1, H_2)
h	Heat of ablation (h_1, h_2, h_3, h_4, h_f, h_p, h_v)
J	Coefficient of ablation (J_0)
j	Luminosity of meteor
K	Coefficient of drag
k_t	Thermal conductivity
k_i	Probability of inelastic transition
M	Mass of meteor or meteoroid (M_1, M_2, M_0)
m	Mass ratio (m_1)
m	Molecular weight (only in Section 3.4)
p	Pressure (p_c)
p_v	Vapor pressure
p_n	Aerodynamic normal pressure
p_s	Drag per cm² of surface; or per cm² of cross section
p_t	Tangential drag
p_r	Drag limit of rupture
Re	Reynolds number
r	Equivalent radius of meteoroid (r_0, r_0'; r_d, r_s, r_f, r_b; Δr)
r_n	Radius of atomic orbital (r_k')
S	Surface area of meteoroid
T	Absolute temperature (T_a, T_c, $\bar{T}$, T_0, T_s, T_f; ΔT_0, ΔT_r, ΔT)
t	Time or time interval (t_0)
U	Total energy transfer to meteoroid

V Volume (Section 4.1)

V Potential energy of atomic interaction (except Section 4.1)

V_0 Energy of interaction at closest approach

w Velocity of meteor (w')

X Parameter of ablation (X_e)

Z Zenithal angle of incidence

α Angle of incidence (α_0)

β Luminous efficiency ($\beta_a, \beta_n; \beta_1, \beta_2, \beta_0; \beta_t, \beta_b; \beta_i, \beta_c, \beta_c'$)

Γ Coefficient in mass—velocity relation

γ Coefficient of net heat transfer to meteoroid ($\gamma_v, \gamma_c \bar{\gamma}, \gamma_0$)

$\mathfrak{a}$ Density of meteoroid

∂ Dilution factor of coma ($\partial_0, \partial_\beta$)

ζ Jet correction

ζK Deceleration factor

η Coefficient of viscosity

κ Coefficient of accommodation ($\kappa_n, \kappa_0, \kappa', \bar{\kappa}, \bar{\kappa}_1$)

λ Half-energy range ($\lambda_c, \lambda', \lambda_1$)

λ_0 Gaskinetic mean free path

μ Molecular or atomic mass (μ_1)

ξ Intensity of evaporation

ρ Density of atmosphere, or medium (ρ_c, ρ_0)

σ Meteoroid cross section ($\bar{\sigma}$)

σ_i Atomic cross section [$\sigma(\Omega_i)$]

τ Target radius

ϕ Atmospheric density factor (ϕ_c, ϕ_f)

χ Coefficient of recombination

Ω Quantal energy created in atomic collision (Ω_i)

$\mathfrak{M}$ Stellar magnitude ($\mathfrak{M}_1, \mathfrak{M}_2$)

$\mathfrak{M}_0$ Absolute magnitude of meteor

References*

1. E. Öpik, *Acta et Comment. Univ. (Dorpat.)* **A 3**, No. 5 (1922); *Tartu Obs. Publ.* **25**, No. 1 (1922).
2. J. Hoppe, *Astron. Nachr.* **262**, 169 (1937).
3. F. A. Lindemann and G. M. B. Dobson, *Proc. Roy. Soc. (London)* **A 102**, 411 (1923).
4. C. M. Sparrow, *Astrophys. J*, **63**, 90 (1926).
5. H. B. Maris, *Terr. Mag.* **34**, 309 (1929).
6. E. Öpik, *Acta et Comment. Univ. Tartu.* **A 33** (1937); *Tartu Obs. Publ.* **29**, No. 5 (1937).
7. E. Öpik, *Acta et Comment. Univ. Tartu. (Dorpat.)* **A 26**, No. 2 (1933); *Harvard Coll. Obs.* Reprint No. 100 (1933).
8. E. J. Öpik, *Mém. Soc. Roy. Sci. Liège* **15**, 125 (1955); *Armagh Obs.* Contrib. No. 14 (1955).
9. E. J. Öpik, *Irish Astron. J.* **3**, 165 (1955); *Armagh Obs.* Leaflet No. 30 (1955).
10. F. L. Whipple, *Bull. Am. Meteorol. Soc.* **33**, 13 (1952)
11. The Rocket Panel, *Phys. Rev.* **88**, 1027 (1952).
12. S. Chapman, *Proc. Roy. Soc. (London)* **A 132**, 353 (1931), *et al.*
13. G. J. Phillips, *J. Atmo. Terr. Phys.* **2**, 141 (1952).
14. J. S. Greenhow, *J. Atmo. Terr. Phys.* **2**, 282 (1952)
15. W. G. Elford and D. S. Robertson, *J. Atmo. Terr. Phys.* **4**, 271 (1954).
16. E. J. Öpik, *Geophys. Bull. Dublin Inst. Adv. Stud.* No. 8 (1953), *et al.*
17. M. Nicolet and F. Mange, "Scientific Report No. 35 of the Ionospheric Research Laboratory of the Pennsylvania State College," 1952.
18. *Rocket Exploration of the Upper Atmosphere*, edited by R. L. F. Boyd and M. J. Seaton, Pergamon Press, London, 1954.
19. J. G. Galle, *Abhandl. Schles. Ges. f. Vaterl. Kultur. Abt. Naturw.*, Breslau, 1868.
20. R. Horowitz and H. E. LaGow, *J. Geophys. Research* **62**, 57 (1957); E. T. Byram, T. A. Chubb, and H. Friedman, *Phys. Rev.* **98**, 1594 (1955); *J. Geophys. Research* **61**, 251 (1956).
21. S. K. Mitra, *The Upper Atmosphere*, 2nd ed., Calcutta, 1952.
22. D. R. Bates and M. Nicolet, *Publs. Astron. Soc. Pacific* **62**, 106 (1950).
23. H. S. W. Massey, *Advances in Phys.* **1**, 395 (1952).
24. C. C. Trowbridge, *Astrophys.* J. **26**, 95 (1907)

* *Note added in proof:* A recent monograph by B. J. Levin, *The Physical Theory of Meteors and Meteoric Matter in the Solar System* (in Russian), Moscow, 1956, 294 pp., was received too late by the author to be included in the list of references.

25. L. Vegard and G. Kvifte, *Nature* **162**, 967 (1948); H. G. Jenkins, J. N. Bowtell, and R. W. Strong, *Nature* **163**, 401 (1949); K. G. Emeléus, N. D. Sayers, and R. A. Bailey, *Nature* **166**, 656 (1950).
26. E. J. Öpik, *Proc. Roy. Soc.* (*London*) A **230**, 463 (1955).
27. M. Plavec, *Bull. Astron. Inst. Czechoslov.* **2**, 19 (1950).
28. E. P. Henderson, Annual Report of the Smithsonian Institution, 1948, p. 257.
29. E. J. Öpik, *Irish Astron. J.* **3**, 206 (1955); *Armagh Obs.* Leaflet No. 39 (1955).
30. E. J. Öpik, *Irish Astron. J.* **4**, 84 (1956); *Armagh Obs.* Contrib. No. 19 (1956).
31. G. Edwards, *Nature* **176**, 109 (1955).
32. A. V. Nielsen, Meddel. *Ole Römer Obs.* No. 17 (1943).
33. W. Wahler, *Geochim. et Cosmochim. Acta* **9**, 105 (1956).
34. F. L. Whipple, *Astrophys. J.* **111**, 375 (1950).
35. P. M. Millman, *Harvard Coll. Obs. Ann.* **82**, 113 (1932); **82**, 149 (1935).
36. P. M. Millman, *Nature* **172**, 853 (1953), *et al.*
37. A. F. Cook and P. M. Millman, *Astrophys. J.* **121**, 250 (1955).
38. E. Öpik, *Tartu Obs. Publ.* **28**, No. 6 (1936).
39. C. C. Van Voorhis and K. T. Compton, *Phys. Rev.* **37**, 1596 (1931).
40. W. E. Forsythe, *Smithsonian Physical Tables*, 9th ed., Washington, D.C., 1954; also F. E. Fowle, 8th ed., 1933.
41. H. S. W. Massey and E. H. S. Burhop, *Electronic and Ionic Impact Phenomena*, Oxford, 1952.
42. E. Öpik, *Harvard Coll. Obs. Ann.* **105**, 549 (1937).
43. E. J. Öpik, *Irish Astron. J.* **4**, 49 (1956); *Armagh Obs.* Leaflet No. 44 (1956).
44. E. J. Öpik, "Statistical Results from the Arizona Expedition for the Study of Meteors," *Armagh Obs.* Contrib. No. 23 (1958), in press.
45. L. G. Jacchia, Z. Kopal, and P. M. Millman, *Astrophys. J.* **111**, 104 (1950).
46. E. J. Öpik, *Irish Astron. J.* **2**, (1953); *Armagh Obs.* Leaflet No. 20 (1953).
47. D. R. Bates and H. S. W. Massey, *Phil. Mag.* **45**, 111 (1954).
48. N. Herlofson, *Repts. Progr. in Phys.* **11**, 444 (1948).
49. J. S. Greenhow an G. S. Hawkins, *Nature* **170**, 355 (1952).
50. A. C. B. Lovell, *Sc. Progr.* **38**, 22 (1950).
51. E. Öpik in "Report on Geophysical Discussion," *Observatory* **69**, 190 (1949).
52. P. M. Millman, *J. Roy. Astron. Soc. Can.* **44**, 209 (1950).
53. F. M. Durbin, *Phys. Rev.* **30**, 844 (1927).
54. T. R. Kaiser and R. L. Closs, *Phil. Mag.* **43**, 1 (1952).
55. T. R. Kaiser, *Phil. Mag. Suppl.* **2**, 493 (1953).
56. J. S. Greenhow and E. L. Neufeld, *J. Atmospheric and Terrest. Phys.* **6**, 133 (1955).
57. R. M. Sutton and J. C. Mouzon, *Phys. Rev.* **37**, 379 (1931).
58. Ph. Rudnick, *Phys. Rev.* **38**, 1342 (1931).
59. D. R. Bates and A. Damgaard, *Trans. Roy. Soc.* (*London*) A **242**, 101 (1949).
60. D. V. Atterton and T. P. Hoar, *Nature* **167**, 602 (1951).

61. F. K. Dalton, *Roy. Astron. Soc. Can.* **44**, 1, 185 (1950).
62. F. A. Kloverstrom and W. A. Rense, *Astrophys. J.* **115**, 495 (1952).
63. J. F. Evernden and J. Verhoogen, *Nature* **178**, 106 (1956).
64. A. F. Cook, *Astrophys. J.* **120**, 572 (1954).

Index

DOVER PHOENIX EDITIONS

An Introduction to the Approximation of Functions, Theodore J. Rivlin. Unabridged republication of the 1969 edition. 160pp. 49554-X

An Essay on the Foundations of Geometry, Bertrand Russell. Unabridged republication of the 1897 edition. 224pp. 49555-8

Elements of Number Theory, I. M. Vinogradov. Unabridged republication of the first edition, 1954. 240pp. 49530-2

Asymptotic Expansions for Ordinary Differential Equations, Wolfgang Wasow. Unabridged republication of the 1976 corrected, slightly enlarged reprint of the original 1965 edition. 384pp. 49518-3

Physics

Semiconductor Statistics, J. S. Blakemore. Unabridged, corrected, and slightly enlarged republication of the 1962 edition. 141 illustrations. xviii+318pp. 49502-7

Wave Propagation in Periodic Structures, L. Brillouin. Unabridged republication of the 1946 edition. 131 illustrations. 272pp. 49556-6

The Conceptual Foundations of the Statistical Approach in Mechanics, Paul and Tatiana Ehrenfest. Unabridged republication of the 1959 edition. 128pp. 49504-3

The Analytical Theory of Heat, Joseph Fourier. Unabridged republication of the 1878 edition. 20 figures. 496pp. 49531-0

States of Matter, David L. Goodstein. Unabridged republication of the 1975 edition. 154 figures. 4 tables. 512pp. 49506-X

The Principles of Mechanics, Heinrich Hertz. Unabridged republication of the 1900 edition. 320pp. 49557-4

Thermodynamics of Small Systems, Terrell L. Hill. Unabridged and corrected republication in one volume of the two-volume edition published in 1963–1964. 32 illustrations. 408pp. 6½ x 9¼. 49509-4

Theoretical Physics, A. S. Kompaneyets. Unabridged republication of the 1961 edition. 56 figures. 592pp. 49532-9

Quantum Mechanics, H. A. Kramers. Unabridged republication of the 1957 edition. 14 figures. 512pp. 49533-7

The Theory of Electrons, H. A. Lorentz. Unabridged reproduction of the 1915 edition. 9 figures. 352pp. 49558-2

The Principles of Physical Optics, Ernst Mach. Unabridged republication of the 1926 edition. 279 figures. 10 portraits. 336pp. 49559-0

The Scientific Papers of James Clerk Maxwell, James Clerk Maxwell. Unabridged republication of the 1890 edition. 197 figures. 39 tables. Total of 1,456pp.
Volume I (640pp.) 49560-4; *Volume II* (816pp.) 49561-2

Vectors and Tensors in Crystallography, Donald E. Sands. Unabridged and corrected republication of the 1982 edition. xviii+228pp. 49516-7

Principles of Mechanics and Dynamics, Sir William (Lord Kelvin) Thompson and Peter Guthrie Tait. Unabridged republication of the 1912 edition. 168 diagrams. Total of 1,088pp. *Volume I* (528pp.) 49562-0; *Volume II* (560pp.) 49563-9

Treatise on Irreversible and Statistical Thermophysics: An Introduction to Nonclassical Thermodynamics, Wolfgang Yourgrau, Alwyn van der Merwe, and Gough Raw. Unabridged, corrected republication of the 1966 edition. xx+268pp. 49519-1

Engineering

Principles of Aeroelasticity, Raymond L. Bisplinghoff and Holt Ashley. Unabridged, corrected republication of the original 1962 edition. xi+527pp. 49500-0

Statics of Deformable Solids, Raymond L. Bisplinghoff, James W. Mar, and Theodore H. H. Pian. Unabridged and corrected Dover republication of the edition published in 1965. 376 illustrations. xii+322pp. 6½ x 9¼. 49501-9